Mirrors and Mazes:

A guide through the

climate change debate

Howard Thomas Brady

Mirrors and Mazes: A guide through the climate change debate

Copyright statement

Other information

Published in Australia in 2017 by Mirrors and Mazes, PO Box 70, Jamison Centre, Canberra, Australia 2614.

1st Edition 2016

2nd Edition 2017

The book is available in Australia from the website
http://www.mirrorsandmazes.com.au

ISBN-10: 1-5466-2911-4
ISBN-13: 978-1-5466-2911-5

A CIP catalogue record of this book has been made with the National Library of Australia.

Designed and typeset by Keep Creative, Canberra, ACT.

Table of Contents

Prologue

In the 1980s and 90s, we thought we knew what the climate was doing: greenhouse gases were in charge.

Then, in the first decade of the 21st century, climate got tricky. Temperatures, albeit at seemingly record levels, paused.

Our earlier ideas were simplistic, our slogans and captions banal, and our computer models mere toys.

So sip the nectar of your choice, suspend your judgement for a few hours and prepare to enter the world of climate.

Wherever we look, we are forced to look elsewhere. When we rush in one direction, we find we have to back-track and find another path.

We really are in a world of mirrors and mazes.

Chapter 1.

The climate debate

Time travel

If we could transpose ourselves to chilly Europe around 360 years ago, what would we have thought as the glaciers advanced and rivers, like the Thames, froze over in the winter months? Would we have panicked?

In June 1644, would we have joined the procession led by the Bishop of Geneva to the Des Bois Glacier near Chamonix to beg God to stop the glacier advancing? Or in 1653, after seeing the advance of the Aletsch Glacier, would we have asked the Jesuits of Siders to do penance? Or in 1670, would we have petitioned the nearby Capuchin monks to pray for the Vernagt Glacier near Vent to stop?

The answers to these questions would be interesting. In truth, the present scientific establishment and media would have imploded as they watched the relentless advance of those glaciers. And what about the media headlines? They would have proclaimed the end of the world was imminent!

For the past few decades, we have been in a state of *panic over global warming*. It is argued that we are now the climate makers living in a totally different regime and facing a dire challenge due to a warming threat brought on by our own industrialisation – a warming that could destroy us. But, is this present warming as *man-made* as we think it is? And are we inexorably marching to self-destruction? These questions are central to the modern climate debate.

The main event

The modern climate debate has been dominated by a United Nations body called the Intergovernmental Panel on Climate Change (IPCC). Its reports cover: particular consensus views on climate science and conservation issues; and strategies by which industrialised nations implement carbon reduction schemes that it believes may curtail global warming.

The Internet has become a significant platform to highlight issues in the climate debate. There is now a plethora of internet sites espousing all sides of this debate. Most of these sites stridently defend a certain point of view and have helped to further polarise the debate. Even the publicity machines of individual universities tout new scientific articles online. Such promotion is used to highlight particular views, to further enhance the profile of their institution and to attract extra financial funding.

Many scientists argue that there is a danger of catastrophic ecological change from rising sea levels and higher temperatures. They are convinced that mankind is responsible and that the future of the world mainly depends on reducing greenhouse gases, especially carbon dioxide.

Other scientists argue that the sensitivity of the climate system to rising carbon dioxide levels has been wildly overestimated and curbing them will do little to halt global warming. These scientists do not dispute that over the past 300 years the Earth has been warming, the sea level rising and the atmosphere experiencing some additional warming from rising carbon dioxide levels. But, for these scientists, such changes are within the parameters of other historical warm periods that lasted 300 to 600 years and are mainly due to natural forces in the Earth-Sun climate system. Their arguments imply that billions of dollars are being wasted in an attempt to curb greenhouse gases.

For those who talk about present temperatures being *unprecedented*, geological history shows that for 80% of Earth's history there have been no ice sheets at the poles. Indeed, Mother Earth has normally been at least 3°C to 5°C warmer, and plants and animals have happily lived and evolved in those hotter conditions.

The world has warmed over the past 300 years. Yet not as much as various warm periods in the last 8000 years. More of our modern thermometer temperature records may be broken, but such records only go back 150 years.

During the warming of the last 300 years there have been periods when temperatures rose while at other times temperature changes paused. However, the general trend has been one of warming. During this entire time carbon dioxide levels increased only in the last 150 years mainly due to mankind's increasing use of fossil fuels. A possible link between rising carbon dioxide levels and temperature seemed to exist in the 23-year period between 1975 and 1998, as both rose seemingly in harmony. But then the game changed as temperatures paused in the past 17 years while carbon dioxide levels rose over 9%. And this recent pause was not an isolated incident. Between 1945 and 1975, carbon dioxide levels rose strongly after World War II, and temperatures paused for 30 years. Another similar period was late in the 19th century. Such patterns are clear even if not properly understood, and they indicate there have been at least 70 years of temperature pauses in the past 150 years while carbon dioxide levels increased.

Observing these temperature pauses, one scientist who changed his views was James Lovelock, a world renowned English environmentalist and founder of the Gaia hypothesis on the interconnectivity of all living and non-living things. Dr James Lovelock was once a climate catastrophist predicting temperature rises of 5°C to 8°C in the 21st century; temperatures that would reduce most arable land to desert. His views, expressed a few years ago, are now very different:

> The problem is we don't know what the climate is doing. We thought we knew 20 years ago. That led to some alarmist books – mine included – because it looked clear-cut, but it hasn't happened. The climate is doing its usual tricks. There's nothing much really happening yet. We were supposed to be halfway toward a frying world by now. The world has not warmed up very much since the millennium. Twelve years is a reasonable time … it (the temperature) has stayed almost constant, whereas it should have been rising – carbon dioxide is rising, no question about that (Lovelock, 2014).

Solutions before answers

The push to curb greenhouse gases fits well with a strong conservationist movement that seeks to curtail coal mining and with some good reasons. The mining of billions of tonnes of coal a year is not sustainable in the long-term. Arable land that could always provide food into the distant future is being ruined, and land mined by large open cut operations cannot be rehabilitated to its former state.

There is no question that we will benefit from economical energy sources other than coal, oil and gas, especially for transport and electricity generation. So there is good justification for the installation of experimental plants of various kinds, but there is no wisdom behind the large scale installation of solar and wind technologies that, due to poor efficiency, are only in the infant stages of their development.

In the fervent rush to reduce greenhouse gases, many countries have installed large-scale solar plants and wind farms that require large subsidies or tax credits to justify some return of capital. In Spain the economy was not robust enough to fund those subsidies that became a significant component in its ballooning national debt.

Warren Buffett, the *prophet of Omaha*, a capitalist hero in the United States has ploughed billions into alternative energy. His investment is $US15 billion dollars and soon to double! He has done the maths. He is attracted to the tax credits as they provide him with offsets against his other businesses making real profits. Who could not but fall in love with an alternative energy bond, especially when the profits are being paid by the taxpayer?

Will this dependence on subsidies compromise a banking system that will then vigorously defend such policies, even if proved ineffective, so banks do not sustain huge write-downs? Will governments keep taking the risk of underwriting private banks if their exposure to risky technology ventures threatens a national banking system?

Comments by Google (obviously convinced by computer modellers that we are facing a climate catastrophe) reflect the truth about the current technologies. Google spent millions of dollars on research and development to use clean energy for its headquarters. It reported to the *Institute of Electrical and Electronic Engineers* (IEEE) in New York:

> ... we had shared the attitude of many stalwart environmentalists: we felt that with steady improvements to today's renewable energy technologies, our society could stave off catastrophic climate change. We now know that to be a false hope ... Renewable energy technologies simply won't work; we need a fundamentally different approach.

The message is clear. Monies spent installing large-scale inefficient technologies are better redirected to fund research, both into new and improved alternative energy technologies and also into designing strategies for better adaptation to climate change. This message was reinforced by billionaire, Bill Gates, the founder of Microsoft, who is prepared to invest further large sums into alternative energy research and development projects. Gates told the Financial Times in June 2015 that the focus should be on increasing research because the current technologies could reduce carbon dioxide emissions at only a '*beyond astronomical cost*'. Gates further commented that:

> There's no battery technology that's even close to allowing us to take all of our energy from renewables and be able to use battery storage in

order to deal not only with the 24-hour cycle but also with long periods of time when it's cloudy and you don't have sun or you don't have wind.

Unfortunately, both solar and wind technologies have to be backed up by fossil fuel generators. Those cheaper coal-powered plants take over 24 hours to become operational and cannot be turned on or off quickly. They must be operational when there is no wind. Even when there is a lot of wind, they must still be left on, ready for use, as wind velocities can drop to zero within minutes. As an example, Ian Plimer, Emeritus Professor of Earth Sciences at the University of Melbourne, and Professor of Geology at the University of Adelaide, Australia, described the situation during heat waves in Australia in January 2014 when the South Australian grid was stretched to its limit of 12,000 megawatts. During that time 28 wind farms with a design capacity of 1200 megawatts were able to only contribute a measly 128 Megawatts to the grid! And despite this anomalous situation, South Australia boasts that 40% of its electricity comes from wind power! So even at times when the wind turbines are delivering power, much of that fossil fuel electrical capacity is still running at additional cost (Plimer, in Moran, 2014).

This means that the large-scale installation of the present alternative technologies achieves very little except in areas where wind and solar are extremely reliable, and even then the cost is questionable.

False prophets

The climate debate itself has been dominated by charismatic prophets who warn of catastrophic climate change. Some have achieved high public status, but their extreme predictions of climate change have not eventuated. They predicted certain dams, now full, would dry-up within a decade. They predicted severe storm frequencies that statistics show have not occurred. And they are still warning of catastrophic sea level rise, despite no evidence it is accelerating (that is, getting faster and faster), but rather steadily rising around 10 cm to 15 cm per century.

Some of these soothsayers have lost, or are fast losing, much of their credibility. Others are simply forgiven and not held to account for their predictions. We would sack a stockbroker who lost us thousands of dollars due to wrong advice, but somehow these *prophets* are allowed to make *little errors* and to simply keep shifting out their doomsday forecasts by a decade or two. These prophets are treated far more kindly than Count Von Iggleheim. This German nobleman followed the ideas of a priest astrologist, Johannes Stoffler, and predicted a biblical flood for February 20, 1524. The crowds gathered, some people boarded his purpose-built three-storey ark, but when the flood did not eventuate, the locals rioted and stoned the Count to death (Doe, 2006).

Anyhow, despite their failed predictions, some of these false prophets are still the *go-to people* if the media wants a dramatic headline. They love simplistic arguments and often stifle debate. They appeal to emotion when the general public needs well-argued science. In their enthusiasm, they use all the climate opposites to rest their case. So each heat wave or cold snap, each drought, flood or bushfire, each increase in pack ice or collapsing ice shelf, is a harbinger of impending doom. So now *single weather events have become synonymous* with *global warming-climate change* and imply *global catastrophe.*

Years ago, during past periods of rising temperatures, such as the Medieval Warm Period (a 1000 years ago), the Roman Warm Period (2000 years ago), or during an even earlier, warmer period known as the Holocene Thermal Maximum (also known as the Holocene Climatic Optimum (4500 to 8000 years ago), any doomsayer could have used successive decades of higher temperatures as added proof of ideas based on religion, mythology, astrology, ouija boards, tea leaves or whatever. Should our present world begin to warm further after the current pause – a scenario with a certain degree of probability – the present climate prophets will shout from the rooftops that such change validates their forecasts for a baked Earth. *This is a devilish argument that can make concurrence of events a proof of anything at all.*

Yet, if the world cools significantly after this current pause, the present prophets will be quickly relegated to anonymity. Other prophets will emerge with forecasts that fit a changing doomsday narrative. Their prophecies will not be of rising sea levels and a baked planet, but of freezing conditions, famine, disease and drought.

The trouble is that climate catastrophists tend to extremes. Their enthusiasm polarises debate and weakens rational argument. Reason and science are clouded by the emotion of enthusiasm, and the influence of religion is easily misused.

Models and history

Underlying the confusion and exaggeration in the climate debate are the scientists playing with their computers and tuning their climate models. Many of the present models have led to panic and poor decisions that threaten economic development and human welfare without providing any future benefit.

There are many computer models trying to mimic the climate so that precautionary scenarios can be developed, and these are referred to as General Circulation Models (GCMs). They involve complex computer programmes with internal variables not very transparent to scrutiny. These models have driven the climate debate for the past 40 years and have been overused as predictive tools. As research aids, they can show only how the inputs *inside* the models react, not how they behave *outside* in the real world. Furthermore, other factors critical in the climate system may remain unknown or not considered.

Historical climate patterns and cycles that have occurred in the past, not just computer models, need to be recognised as important predictive tools, even if those climate patterns and cycles defy explanation. For instance, we know that millions of years ago, on multiple occasions, there were ice ages when carbon dioxide levels were many times higher than those of today. That information tells us other climate forces can override any warming effect of carbon dioxide. We also know that in the past 10,000 years there were abrupt climate changes that occurred quickly (that is, inside decades), and these are still not properly understood. Even in the past 400 years, there was a concurrence of low sunspot numbers with cool periods on two occasions. History must be recognised and allowed to speak as such historical patterns have occurred and may happen again; even if we do not quite understand what triggers them.

Climate is a complex topic embracing many disciplines such as atmospheric physics, meteorology, solar and planetary physics, geological history, oceanography, ice sheet dynamics, plant ecology and animal ecology. We should not expect too much of the models, no matter how powerful our computers. We are grappling with a tricky, nonlinear, chaotic maze that stretches throughout space and in time. Nothing is ever still. The momentum of millions of climatic variables at any particular point of time cannot be expressed by sets of static numbers derived by man.

Pivotal questions

So, is the link between rising greenhouse gases and warming all it is cracked up to be? There are questions that need answers:

- Carbon dioxide levels rose about 11% between 1975 and 1998 when temperatures were also increasing. However, carbon dioxide levels rose another 9% between 1998 and 2017, while temperatures did not increase. So how can rising carbon dioxide levels be the main driver?

- The rate of temperature rise was the same for the periods between 1860 and 1880, between 1910 and 1945, and between 1975 and 1998. Yet carbon dioxide levels increased at very different rates during all these periods, 0.7%, 1.0% and 5% per decade, respectively. If carbon dioxide is the controlling factor in any warming period, should not the rate of temperature rise have steepened with the increasing changes in carbon dioxide levels?

- If carbon dioxide is the main driver of recent climate change, why is it that in over 70 of the past 155 years average global temperatures paused as carbon dioxide levels rose?

- Why is it that computer models did not predict the pause in world temperatures after the year 1998, one that is still ongoing in 2017? After all, there is clear evidence of other pauses in the past 150 years that the computer programmes should have been able to replicate.

- Why is sea level continuing to rise steadily at 10 to 15 cm/century and not accelerating as predicted by the IPCC?

- Why is the public misinformed; being told that due to global warming severe weather events, such as hurricanes, cyclones, tornadoes, are increasing in frequency and getting worse, when this narrative is not supported by worldwide weather records?

A narrow debate

There is another problem in the whole climate debate; the obsessive focus on global warming to the exclusion of global cooling. Independent of any views on global warming, whether from greenhouse gases, or changes in the Sun's magnetic field, or a mix of many factors, *there are now numerous peer-reviewed scientific articles arguing for global cooling within the next century*. These articles are from scientific departments all over the world. Some are based on the recent behaviour of the Sun, whereas others are focused on the role of cosmic rays and cloud formation. While there are variations in argument, the overall consensus in these articles is that there is a high probability of global cooling. Those studying sunspot cycles argue for global cooling within the next 40 years. Those looking at long-term cosmic radiation patterns are less precise, but argue for cooling within the next 100 or so years.

The United Nations has taken on a worldwide role of promoting responses to global threats such as poverty, disease, or climate. Yet there has been no coordinated policy within the IPCC to consider the possibility of global cooling. Again, this should not be turned into another *doomsday* scenario by a new gaggle of prophets, as such a possibility could be mitigated by forward planning. Cooling means some shifts in agricultural belts, some significant shifts in weather patterns; changes that in the past led to huge migrations. Modern society with better developed technology is far more adaptable, but cooling could put certain regions in the world under considerable stress.

This lack of a balanced approach to climate change highlights the political, non-objective nature of the present structure of the IPCC. Sadly a lobby of scientists with catastrophic views of global warming and sea level rise due to rising carbon dioxide levels now control the IPCC. With the population of the globe moving towards 10 billion, adaptation to global cooling, as well as to global warming, should both be considered at the highest levels by the United Nations.

It is therefore critical to explore climate change in all its intricacy. There are so many forces at work that fixation on one possible driver of climate change runs the danger of oversimplifying and distorting our understanding of a rich and complex system.

We really are exploring in a world of Mirrors and Mazes.

Chapter 2.

Climate and chaos

What is linked with what?

The modern climate debate is about the extent that rising carbon dioxide levels are influencing world temperature through what is called the *Greenhouse Effect*. This process involves: shortwave radiation from the Sun warming the Earth's surface; the Earth's surface, in turn, emitting some longwave radiation; various gases, including carbon dioxide, trapping certain wavelengths of that radiation; and then those gases sending some of that radiation back to warm the Earth's surface and atmosphere.

The link between carbon dioxide and temperature is often described as a simple cause and effect relationship – much like a pool cue and a cue ball where carbon dioxide (the cue) simply causes temperature (the ball) to move along. This type of relationship is linear as it is easy to relate the output (how far the ball moved) back to the input (the speed and force of the cue striking the ball).

However, *the climate system is vast with multiple forces all interacting with each other.* These forces are large in number, and, to enhance the complexity, they also operate through different timescales and in different, even over-lapping spaces. So, like the cogs of a bicycle, the influence of these forces ranges from those with shorter cycles with smaller effects, such as diurnal or lunar cycles, to those with longer cycles with larger effects, such as planetary or even galactic cycles.

In such a vast system simply attributing recent global warming to increasing carbon dioxide levels is problematic. When there are multiple causes and complex effects, the final result (the output) cannot be simply related back to the various inputs. In addition, *the minutest change in the initial conditions can also result in diverse, even opposite outcomes because nonlinear, chaotic behaviour can also occur.*

Chaos or nonlinear theory

A branch of science called *Chaos Theory or Nonlinear Theory*, has developed over the past 50 years to further our understanding of complex systems that cannot be reduced to simple relationships between a few variables. The pioneer of nonlinear theory was Emeritus Professor Edward Norton Lorenz who was a mathematician and meteorologist at the Massachusetts Institute of Technology. His seminal paper entitled *Deterministic Non-periodic Flow* was published in the *Journal of Atmospheric Physics* in 1963. Nonlinear theory is now widely used and has led to a better understanding of countless diverse problems; for example, unstable heart rhythms (arrhythmia) or population dynamics or instability in electrical grids.

Take our own human system. There are myriads of interacting chemicals and complex electrical pulses. There are the blood, lymph and nerve networks, and then the sleep-wake, the male-female hormonal, and aging cycles. Each organ has its own functions and rhythms. There are different temperatures in various organs or parts of the body. Body temperature varies whether one is hungry, sleepy, cold, undertaking strenuous activity or digesting food. Temperatures vary in female hormonal cycles. The body also tries to control temperature through the pores of the skin, through breathing and by modifying its own environment through clothing. The tipping points in the body's internal temperature system leading to death, are hyperthermia (too hot, around 45°C) or severe hypothermia (too cold around 24° to 28°C).

Sometimes a doctor can explain what is happening. Often various explanations for a certain condition or event can turn out to be wrong or inexplicable; often any number of apparent unrelated factors could each cause the same condition. More importantly, amid all this complexity, the same initial conditions do not always lead to identical outcomes. That is due to the ultimate quirkiness of a nonlinear system.

Detectives at a crime scene gather a body of evidence and then work backwards to establish the history of a crime. But, in a nonlinear system, it may be impossible to work backwards from a particular outcome to define its actual cause. It is not just the complexity of the interrelationships between all the variables. *The nonlinearity of the system means that different outcomes, including the outcome of nothing happening at all, can actually result from the same conditions!*

Balance, stability and tipping points

Paradoxically, nonlinear systems can develop a balance and a stability that make it extremely difficult, without some totally outside event, to go beyond certain boundaries. The more forcings, the more inter-relationships, the more interactions, *BUT the more stability*. So, natural checks and balances can keep a system such as climate within boundaries even though there is a certain degree of unpredictability within those boundaries. There are limits to how cold it can get at a certain place, or how windy, or how wet, or how humid, but the random variations in temperature, or wind, or rain, or humidity within those limits are so frustrating that, even with modern computers, *no one can predict with certainty climate-weather in a month's time.*

Scientific research may identify more and more patterns in climate that provide better predictive tools for the future and more accurate knowledge of possible interactions. But, nonlinear behaviour is the most difficult to model, as different outcomes can result from the same parameters, or the smallest variations in the initial conditions. This is often referred to as the *butterfly effect* on the basis that when a butterfly flaps its wings, it could affect a hurricane on the other side of the world.

There is argument that warming caused by rising greenhouse gases will raise global temperatures with catastrophic effects, so Earth is on the edge of a climate precipice, or what is called a *tipping point*. But, the unpredictability of a nonlinear system does not mean it is particularly easy to tip the system out of balance; quite the opposite. Professor Wallace Broecker at Columbia University's Lamont Doherty Earth Observatory describes nonlinear stability when writing about the turning points of recent ice ages:

> ... the spacing between these terminations was never exactly 100,000 years. Rather it was either close to 80,000 years or close to 120,000 years. This gives the impression that during each glacial episode the Earth system drifted toward some sort of instability that, when reached, triggered a jump back to a warm state (Broecker, The Great Ocean Conveyor 2010, p.11).

In other words, nonlinear systems bounce away from catastrophe. Extraordinary events are needed to tip these systems out of balance. In the case of the Earth's climate system the coupling of the Earth's atmosphere and oceans has led to a state that has been stable enough over the last four billion years for life to prosper despite huge changes in greenhouse gas levels, sea levels, ocean volumes, ice volumes and local climates.

Geological history indicates that other causes have often overridden any effect of greenhouse concentrations many times higher than those of today or even those envisaged over the coming centuries. There have been estimates by various scientists of carbon dioxide and temperature levels in the past, for example by Dr Chris Scotese of the University of Texas (1999) and Professor Robert Berner of Yale University (1990). There was an ice age 450 million years ago with carbon dioxide levels 10–15 times higher than today. There was also another ice age 350 million years ago when carbon dioxide levels were as low as today (Diagram 2.1).

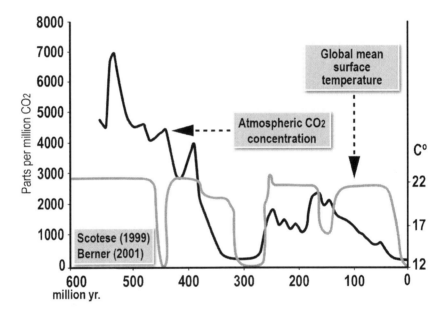

Diagram 2.1: Temperature and carbon dioxide levels over the past 600 million years (based on Scotese, 1999 and Berners, 1990).

Ice-age conditions occur with widely different carbon dioxide levels. Our present carbon dioxide levels are around 400 parts per million. The ice age 430 million years ago had carbon dioxide levels around 4500 parts per million; the ice age around 300 million years ago had carbon dioxide levels close to present levels; the ice age around 140 million years ago had levels around 2000 parts per million. These variations clearly show other forces can override the greenhouse effect of very high carbon dioxide levels.

There are times when carbon dioxide and temperature seemed in sync, but at other times they moved in opposite directions. The present carbon dioxide levels are low from a historical perspective and the recent increase in greenhouse gases is not unprecedented, but well within their long-term variability. The fact that carbon dioxide and temperature are often out of sync is clear evidence that many other factors are driving the climate system and we need to understand climate as a complex, nonlinear system. History tells us that we cannot simply equate rising carbon dioxide levels with higher global temperatures.

In awe

Climate history is the end-result of the nonlinearity of the climate system. Historical patterns contain the full nonlinearity of that system. The words of Professor Wallace Broecker when he received the Crafoord Prize in 2006 from the Swedish Academy of Science are very apt:

My lifetime study of the Earth's climate system has humbled me. I am convinced that we have greatly underestimated the complexity of the system.

Chapter 3.

Wheels within wheels

Cycles everywhere

In many nonlinear or chaotic systems, such as climate, there are recognisable cycles, events that repeat themselves over time. Such cycles relating to climate may be likened to the layers of an onion, or encased Babushka Russian dolls. Many are even more complex than that as they operate on different timespans and over varying areas of space.

The Grand Ice Ages

During most of the Earth's history there have been no icecaps at the poles and the Earth's surface temperature has been warmer by at least 3°C to 5°C. Animals and plants have happily lived and evolved in these warmer Earth surface conditions. Dinosaurs have sunbaked in Antarctica! The biggest breaks in these hotter Earth surface conditions have occurred due to climate cycles called the *Grand Ice Ages*. They can last up to 30 million years. Each ice age is characterised by a series of ice advances into the Earth's mid-latitudes, and between these ice advances are shorter warm periods called interglacial periods. We are presently in a warm period within one of these *Grand Ice Ages*. The most recent ice advance concluded about 12,000 years ago. Even though we are warmer than during the ice advance 80,000 years ago, the present climate on Earth is colder than our planet's average climate over the past four billion years.

Why the cyclicity? The solar system consists of various planets, asteroids and comets that orbit around the Sun, trapped by its gravity. However, in turn, the solar system is like a pilgrim on a long journey within the Milky Way galaxy where each pilgrimage or orbit lasts millions of years.

Astrophysicist Nir Shaviv at the Racah Institute of Physics at the Hebrew University of Jerusalem has noted that major ice ages on Earth coincide with periods when the solar system crosses through the spiral arms of the Milky Way galaxy during the last 2 billion years (Shaviv, 2003). His results depend on the analysis of the dating and then the changes left in iron meteorites by cosmic radiation. Such transits occur every 143 +/ 25 million years. Shaviv proposes a theory, shown in Diagram 3.1, that when our solar system crosses such an arm, there is a natural increase in high energy cosmic radiation from thousands of exploding stars in each spiral arm. On entering the Earth's atmosphere these highly charged particles, travelling near the speed of light, collide with the gases in the Earth's surface atmosphere. Shaviv argues these collisions create charged atomic particles around which water moisture can condense to form clouds. Since clouds can reflect solar radiation before it reaches the Earth's surface, *more cosmic ray collisions mean an increase in cloud cover and a colder Earth.*

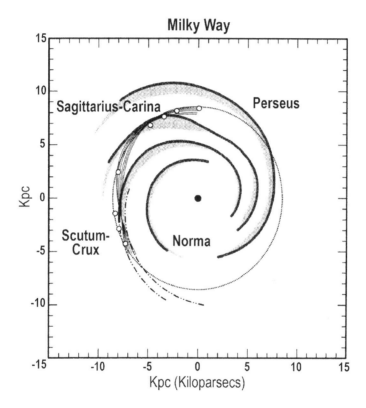

Diagram 3.1: A diagram showing the orbital path of the solar system through the Milky Way (A. Ollila, website, www.ClimatExam.com).

The solar system's orbit crosses through the arms of the Milky Way. The scale is in kiloparsecs, an astronomical measurement, where one parsec is about 3.26 light years, or about 31 trillion kilometres. Shaviv's theory links each transit of an arm of the Milky Way with higher cosmic ray radiation due to the higher number of nearby stars. This radiation increases cloudiness, that in turn, reflects more of the Sun's incoming radiation into space; so significant increases in cloud put the Earth into an ice-house condition.

A world famous isotope chemist Jan Veizer, working in Ottawa, Canada, and independently from Nir Shaviv in Jerusalem, plotted past ocean temperatures using different forms of the oxygen atom in 4,500 calcium carbonate shells of marine microfossils (foraminifera), brachiopods (marine animals with bivalve shells, like clams) or belemnites (an extinct animal shaped much like a cuttlefish) (Veizer, 2005).

Veizer found that periods of cold ocean temperatures closely coincided with Shaviv's transits of the solar system through various arms of the Milky Way. Here was an independent verification of a pattern for the past 500 million years. The absence of preserved fossils older than this that could be analysed with his methods meant Veizer could not extend his analysis as far back in time as Shaviv's analysis of meteorites. However, apart from the transits across the arms of the Milky Way the data also suggested the solar system, as it moved through space, oscillated vertically upwards and downwards across the plane of the Milky Way every 32 million years.

In 2014 Shaviv, Veizer and Andreas Prokoph (another Professor at Ottawa) announced further results for the past 500 million years, now based on 24,000 fossil analyses (Shaviv et. al, 2014). This more detailed analysis clearly verified variations in the Earth's ocean temperature as the solar system oscillated upwards and downwards across the plane of the Milky Way every 32 million years. This pattern, with a full cycle of 65 million years, was recently analysed by Australian scientists. Professor Robert Baker and Professor Peter Flood of the University of New England also noted that other scientists have plotted fluctuations in marine biodiversity, volcanic activity, and sedimentary cycles with the same periodicity (Baker and Flood, 2015).

Ice sorties

Within the recent Grand Ice Age there have been many advances and retreats of ice, each lasting around 100,000 years. These have cyclical patterns that were unravelled a hundred years ago by two extraordinary scientists. Milutin Milankovitch was a Serbian mathematician who developed many of his ideas while a prisoner during World War I (he was captured on his honeymoon!). He was later a professor at the University of Belgrade. The other was the self-educated Scottish genius, James Croll, who taught himself physics and astronomy, and who developed his ideas while working as a janitor at the Andersonian College and Museum at Glasgow. These two men showed how these advances of ice were regulated by changes in the Earth's distance from the Sun, by a wobble in the tilt of the Earth, and by the rotation of this tilt through 360 degrees.

The Earth rotates in an elliptical path around the Sun every year. At the far end of the ellipse the distance to the Sun is 152.17 million kilometres and in the narrow section around 147.2 million kilometres. This is nearly a circular path and the seasonal difference between the maximum and minimum radiation received by the Earth is only around 6%. However, over time the Earth's orbit becomes more elliptical in 95,000–125,000 year cycles. When most out of shape, the difference between the summer and winter radiation received from the Sun, as the Earth travels this more egg-shaped path, is well over 20%.

There is a second cycle because the Earth is not spinning at right angles to the Sun. The spin is tilted and wobbles between 22.1 and 24.5 degrees every 41,000 years. At higher latitudes, close to the polar regions, even a 2.5 degree wobble in the Earth's surface tilt is equivalent to a change in latitude of 2.5 degrees, and would, for example, shift the wheat belts of Canada and Belarus at least a further 300 km to the south. Finally there is a third cycle every 26,000 years as the tilt rotates through 360 degrees. During this cycle winter shifts to December in the Southern Hemisphere, and summer shifts to December in the Northern Hemisphere; and vice versa as the cycle repeats itself.

For Milankovitch and Croll, ice ages start in the high mid-latitudes of the Northern Hemisphere; these areas, north of 60°N, are most susceptible to changes in snow and ice cover. Here, if the climate cools so that winter snow does not melt in summer, then the cold summers would allow snow to remain and accumulate and ice sheets to grow. Conversely, during a sustained period of continued warm summers, as the Earth's surface orbit becomes more circular, these mid-latitude ice sheets melt and collapse, and eventually summers are snow free.

Some smaller cycles

On top of these larger cycles there are many other smaller cycles that can influence climate.

The Sun wobble

The Sun wobbles in a complex series of movements in a pattern that repeats every 179 years. These variations are due to the combined gravitational pull of the planets on the Sun, with Jupiter and Saturn exerting the largest influence due to their size. Such wobbles vary the Earth's surface distance from the Sun, on top of (as discussed previously) the main distance variations due to the Earth's elliptical path around the Sun.

Sunspots

Sunspots are lower temperature areas linked to the Sun's magnetic field that appear and disappear in cycles on the Sun's surface. The number of sunspots changes in a wave-like pattern throughout each cycle that vary in length from 9 to 13.7 years. The highest number of sunspots occurs in the middle of each cycle and the Sun's magnetic field is strongest at this time. *These cycles have been noted since the 17th century and carefully recorded since 1775.* There is a definite correlation between low sunspot cycles and cool periods on the Earth. The exact reasons for this correlation are a subject of debate.

During a sunspot cycle, sunspots start to appear and increase in number and they move around between 35 and 25 degrees north or south of the Sun's equator. Once one cycle finishes in say the northern hemisphere of the Sun, the cycle repeats itself in the southern hemisphere of the Sun. This means the magnetic poles of the Sun continually flip as the cycles repeat. The 9 to 13.7 year cycle is called the *Schwabe cycle*, and the repetition of the cycle in both hemispheres is called the *Hale cycle*. Apart from these cycles, there are other longer sunspot cycles such as the 87 year *Gleissberg cycle*, the 210 year *De Vries Suess cycle*, and the 2300 year *Hallstatt cycle*.

1500 year cycles

There is frequent discussion of 1500±500 year climate cycles named after a Danish scientist, Professor Willi Dansgaard from the University of Copenhagen and a Swiss scientist, Professor Hans Oeschger from the University of Bern, Switzerland. Their work on Greenland ice cores led to the discovery of many cold-warm climatic cycles during the past 250,000 years. Further studies of Antarctic ice cores and ocean sediments have extended the history of such cycles back more than one million years. The more recent warm periods between these cold cycles are the Minoan, Roman and Medieval Warm Periods and, according to many, the present warm period.

In the cooler periods, between these warm cycles, there are sometimes very cold events when armadas of icebergs float down from the Arctic into the Atlantic Ocean, and, as they melt, they dump rock from Eastern Canada, Greenland and other Arctic islands onto the ocean floor, even as far south as north-west Africa. These events in the cold *Dansgaard-Oeschger* cycles are known as *Heinrich events*. There has to be some relationship to the Sun as some rocks dumped by these icebergs contain evidence left by cosmic ray radiation. The weaker the Sun's magnetic field, the more cosmic radiation from high energy stars and even galaxies can reach the Earth. Certain isotopes of beryllium and carbon, for example, can be formed in rocks only by such high velocity radiation and these isotopes are a measure of the cosmic radiation reaching the Earth.

The great ocean conveyor belt

Other cyclical patterns that influence the world's weather and long-term climate change are related to a giant conveyor-belt system that connects all the oceans to an ocean current that flows around Antarctica. Since the upper few metres of the ocean retain the same amount of heat as the entire atmosphere, the ocean, as a whole, can have a huge cooling or heating effect on the atmosphere and so affect air temperature, air moisture and wind velocities. These heat exchanges are energy transfers between the atmosphere and huge ocean currents, thousand of kilometres long. These currents slide over and under each other in an ocean whose average depth is around four kilometres, and each has its own signatures of salinity and temperature.

A famous ocean current is the Gulf Stream that flows north-east from the Caribbean across the Atlantic towards Norway and brings warmer conditions to northern Europe at the same latitude as frigid Hudson Bay in Canada. Then there is what is called the Deep Water. This is the deepest ocean water in the Southern Hemisphere and in some of the Northern Hemisphere. As pack ice forms around Antarctica, a little extra salt is added to the underlying ocean water that, being heavier, sinks to the sea floor and migrates at snail-pace from the polar regions towards the equator and beyond. *This Deep Water, rich in oxygen, often takes over a decade just to reach the equator, but still accounts for more than half of the water in the world's oceans.*

Ocean-atmospheric cycles

There are many complicated climate patterns due to interactions between surface ocean currents and pressure systems in the Earth's atmosphere. Some have been known for centuries, but not understood.

One famous cycle is the El Nino–Southern Oscillation (ENSO) in the tropical and southern Pacific Ocean. This had been noted by Peruvian fishermen. In El Nino phases there are warmer waters in the Central and Eastern Pacific along the south American coast as westerly winds pile waters on this eastern side. Off Peru this warm water means less vertical mixing in the ocean and poor fishing conditions. At the same time along the eastern Australian coast there is an upwelling of colder waters that means reduced rainfall and sometimes drought. In the La Nina phases, the cycles reverse.

Another cycle is the Pacific Decadal Oscillation that is reflected by a switch of surface temperatures between the western and eastern North Pacific Ocean over a 60-year period. In the eastern North Pacific Ocean the cool phase is reflected in high salmon numbers. In one phase of this cycle warm water can flow north through the Bering Strait between Alaska and Siberia and decrease the surface extent of Arctic pack-ice.

Another pattern which has influenced the history of countries on both sides of the Atlantic for thousands of years is the North Atlantic Oscillation. Here a low pressure system over Iceland and a high pressure system over the Azores off the North-African coast switch around. The first phase brings strong westerlies, rain and warmth to Europe. The second or low phase brings cold Arctic winds to Europe and warmer conditions to Greenland.

There are many other oceanic and atmospheric cycles that are tracked by meteorologists. They normally involve east-west or north-south changes in oceanic or atmospheric conditions. These are usually named as to their general location and include such cycles as the Indian Ocean Dipole, the Atlantic Meridional Oscillation, the Arctic Dipole Anomaly, the Antarctic Oscillation, the Arctic Oscillation, and the Pacific North American Flow Pattern.

Cycles and blind alleys

With all of these cycles in play, relating climate over time to any one cycle during any period of the Earth's history can be problematic. Some long-term relationships between the Sun and the Earth's land, oceans and atmosphere can be related to certain climatic patterns, but that does not mean these patterns are understood. Some cycles could be effects of changes that have occurred over many centuries. Concurrence may be fortuitous. Indeed, some changes may be just due to the natural variability and quirkiness in complex nonlinear/chaotic systems:

> ... Natural climate excursions may be much larger than we imagine. So large, perhaps, that they may render insignificant the changes, human-induced or otherwise, observed during the last Century (Cohn and Lins, 2005).

Some cycles seemingly related to each other might be driven by other unknown forces and some cycles could even be resonating in their length with other drivers in the climate system and, other than that, have little significance at all. The idea of resonance connecting various climate cycles has been taken up by physicists, such as Professor Judith Curry and Dr Marcia Wyatt of the Georgia Institute of Technology and Professor

Andrey Kravtsov of New Mexico State University, using a *Stadium Wave* model hypothesis (Curry and Wyatt, 2013, Kravtsov et al., 2014):

> The 'stadium-wave' signal propagates like the cheer at sporting events whereby sections of sports fans seated in a stadium stand and sit as a 'wave' propagates through the audience. In like manner, the 'stadium wave' climate signal propagates across the Northern Hemisphere through a network of ocean, ice, and atmospheric circulation regimes that self-organize into a collective tempo. The stadium wave hypothesis provides a plausible explanation for the hiatus in warming and helps explain why climate models did not predict this hiatus. Further, the new hypothesis suggests how long the hiatus might last ... The stadium wave signal predicts that the current pause in global warming could extend into the 2030s (Press Release-Georgia Tech October 10, 2013).

The Wyatt and Curry hypothesis is an attempt by scientists to describe the net effect of a myriad of patterns (like wheels within wheels) that can be identified as being related to climate.

However, even cycles with a similar periodicity may not be related as to cause and effect. Nonlinear science suggests that diverse forces can have similar effects and that some apparent causes, seemingly driving the system in a certain direction, may not be critical drivers of the system at all. There could be any number of dead ends in such a nonlinear maze.

A famous medical example of a mistaken link was the long held belief that stomach acids were the main cause of peptic ulcers. That theory was eventually changed when two Australian researchers found that bacteria, such as *helicobacter pylori*, could live in the acid conditions of the human stomach and were often a primary cause of peptic ulcers.

Finally, because of the nonlinear nature of the forcings affecting climate, it should be noted that even repeats of the same cycle are not identical. Each wobble of the axis of a planet or the Sun, each glacial or interglacial epoch, each sunspot cycle, each El Nino or La Nina cycle, each short warming trend, each movement of a deep ocean current and each weather system that passes over our house, has its own uniqueness and uncertainties.

There is a parallel with Plato's concept of a *Universal Idea*, where any universal idea, such as *tree*, reflects a myriad of widely different objects, each with its own *singularity*. In this case, the word *cycle* certainly covers a complex variety of forces and events, that are never exactly the same, and that give shape to the Earth's climate system. It follows that the recent increase in greenhouse gases, and their possible influence on world climate, have to be understood within the framework of all the cycles and all the factors that influence this system; a framework we only partially understand.

Chapter 4.

The warming stairway

The mysterious pattern

Scientists began to graph world temperatures in detail in the 19th century as more countries began to collect regular temperature data using mercury thermometers. Two things became clearer as more data were collected: temperature was steadily rising; and this ascension was punctuated by three periods where the temperatures remained level before rising again. The approximate sequence was: 1860–1880 (warming), 1880–1910 (pausing), 1910–1940 (warming), 1940–1975 (pausing), 1975–1998 (warming), 1998–2017- (pausing). These periods reflect a 50 to 60 year cycle in which the periods of warming and pausing are not of equal length, as each varies between 20 and 35 years. It has been like a football match with three periods of active competition with rests in-between.

There is too much cyclicity in these repeating 60-year warm-pause cycles to say they are just by-products of chaotic variability. It is possible these occurred in the past, but we do not have enough resolution in past historical data to track such cycles. Similar cyclicity was even noted in the 18th century from an analysis of an ice core taken in West Antarctica covering the period from 1705 to 2009 (Thomas et al., 2013).

Even though this pattern is described in the earliest IPCC reports, it remains an enigma. There have been various attempts to deny such a 60-year pattern exists. The closest similar pattern within the Earth's own climate system is a switch in ocean temperatures over a 60 year period between the west and east North Pacific Ocean called the Pacific Decadal Oscillation; each switch lasts about 30 years. In the western Pacific Ocean the cool phase is reflected in high salmon numbers. In one phase warm water can flow north through the Bering Strait between Alaska and Siberia and decrease the extent of Arctic pack-ice.

Why the same gradient?

The temperature records for the past 150 years show that during three warming periods since 1860 the temperature gradients were similar. These gradients are expressed as the average temperature rise per decade. Professor Phil Jones of the University of East Anglia calculated the gradient to be around 0.16°C/decade for all three warming periods using data collected by the British Met Office Hadley Centre (BBC News February 13, 2010). Professor Michael Asten of Monash University, Melbourne calculated it to be 0.19°C per decade using NASA data for the two most recent warming periods (pers. comm. – Diagram 4.1).

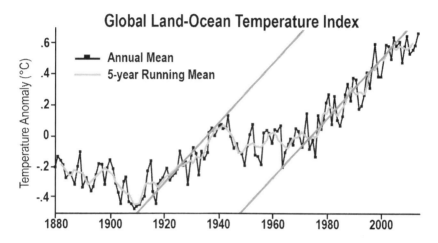

Diagram 4.1: NASA graph of temperature between 1880 and 2010
The gradient lines inserted by Professor M. Asten of Monash University, Melbourne, Australia for the warming periods from 1910 to 1940 and from 1975 to 2000.

Temperature gradients in warming periods for the past 150 years are similar. This is evidence that global warming in each warming period is not accelerating. Similar graphs have been produced by Professor Phil Jones from the University of East Anglia, England. He also shows there was a similar gradient in a warming period between 1860 and 1880.

While the temperature gradients in these warming periods are similar, the changes in carbon dioxide levels are very different. For example, between 1910 and 1945 the rise in carbon dioxide was around 3.3% (or approximately 0.95% per decade), and between 1975 and 1998, the rise in carbon dioxide was around 11% (or approximately 4.8% per decade).

There is also no correlation between rising carbon dioxide levels between periods when temperatures stayed level for a number of years. The rise in carbon dioxide was around 3.4% when temperatures paused between 1880 and 1910, yet during the temperature pause between 1945 and 1975, the rise in carbon dioxide was around 6.5%. Then, during the temperature pause between 1998 and 2017, the rise in carbon dioxide was around 9.0% to 2017 and is still ongoing.

Despite these steady warming gradients, some computer modellers predict a significant increase in temperature this century that requires temperature gradients to steepen sharply. For example, Dr Dan Rowlands at Oxford University and 20 other scientists predict a temperature rise between 1.4°C and 3°C as early as 2050 (Rowlands et al., 2012). How can this scenario be taken seriously when there has already been a 19-year pause in global temperatures since 1998, and when the temperature rise in the past 100 years was only around 0.7°C?

Excuses

A number of scientists try to get around these steady warming patterns by attributing the first and second warm periods in the 19th and 20th century to normal climatic variation, and the third warming period in the late 20th century to rising greenhouse gases from industrial activity. Yet, this explanation is illogical:

- If the warming gradients have been similar in all warm periods since 1860, then there is no logic in saying the recent warming is different or anthropogenic while the other two warming periods are due to natural causes.

- According to standard physics, *the greenhouse warming effect of carbon dioxide behaves in a logarithmic fashion so that any doubling has the same effect* whether it be from 10 parts per million to 20 parts per million or 1000 parts per million to 2000 parts per million. Around the year 1850 when carbon dioxide levels were around 280 parts per million, the warming effect of one extra part per million carbon dioxide was equivalent to around 1.4 parts per million today when carbon dioxide levels are around 400 parts per million. So if you are in the scientific camp that has carbon dioxide as the principal driver of global warming isn't it absurd logic to say that the most recent warming was due to greenhouse gases, but the warming in the mid 19th century was natural. After all, carbon dioxide was more potent in 1850, as an increase of 1 part per million in 1850 was equivalent to 1.4 parts per million today!

- If carbon dioxide levels are such an important driver of world temperatures and are now rising more steeply (as they are) how can there be pauses in global temperatures? Would it not have been easier to have pauses in global temperature in the past when rises in carbon dioxide levels were not so steep?

Best type fits

Some scientists are beginning to incorporate an *oscillating warming trend* in their models. Professor Syun-ichi Akosofu, an astrophysicist from the University of Alaska, has developed a hypothetical model that shows a steady upward-warming trend with regular pauses. Akosofu attributes the oscillations to regular (multi-decadal) heat transfers between the atmosphere and the ocean that are super-imposed over an underlying long-term warming trend. While there are short-term waves of acceleration and deceleration, the long-term, upward sloping temperature gradient is steady (Akosofu, 2010, 2013, see Diagram 4.2). Nicola Scafetta, an Italian, and now a professor at Duke University, is another scientist who also recognises the existence of a step-wise temperature pattern. His work examines possible harmonic patterns due to the interaction of various solar and planetary cycles. His calculations are too mathematically complex to be described here (Scafetta 2012, 2013).

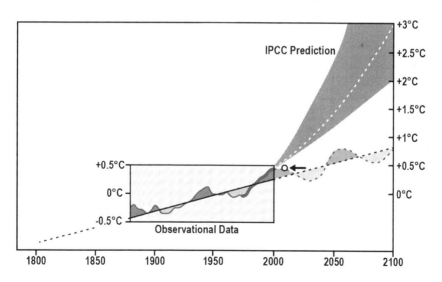

Diagram 4.2: Temperature trends 1880–2008 and then extrapolated (Akosofu 2009).

Akosofu noted an underlying long-term warming trend around 0.05°C per decade. He then superimposed temperature gains and losses when the ocean acted as a heat sink or alternatively exchanged heat with the atmosphere. These gains and losses are calculated as +0.07°C per decade in 30 years of warming and –0.07°C per decade for the other 30 years in a 60-year cycle. The end-result is a slightly negative trend of 0.02°C/decade every 30 years and a positive trend of 0.12°C/decade for the other 30 years of the cycle. The result is an oscillating graph, but there is still an underlying steady warming trend over the past 150 years.

Dr Petr Chylek, from the Los Alamos National Laboratory, and others point out that the high latitude temperature records from Godthaab and Egedesminde in Western Greenland show cooling in the period 1880 to 1915, warming to 1930, cooling to 1990, and warming after 1990 (Chylek et al., 2006, 2007). While these temperature oscillations may differ slightly from those described by Akosofu and Scafetta they still show an underlying oscillating pattern contrary to simplistic climate models.

While Akosofu and Scafetta have different theories, at least they are attempting to find empirical best-fits with 150 years of temperature data. *In contrast, most climate models used by the IPCC are an extrapolation of the warming trend between 1975 and 1998, which does not provide the full picture of what is happening.* Scafetta is very critical of this short-term trendology.

Such short-term trendology has erroneously been used before. Back in the 1960s scientists reacted to a drop in temperature during the temperature pause that had started in 1945. During this time there were predictions of a catastrophic cooling and of an imminent ice age promoted in popular writings such as: *The Climatic Threat. What's wrong with our weather?* (Gribbin, 1978), *Climate and the affairs of men* (Winkless and Browning, 1975), *The weather conspiracy: the coming of a new Ice Age* (The Impact Team, 1977).

Here are extracts from two of those books:

The next 60 years on Earth will be disastrous as the sub-Sahara and Canada experience continued drought and increasing cold, and as the Arctic tundra continues its advance over the wheat growing land of Russia (Climate and the affairs of men, Winkless and Browning, 1975).

... That is the consensus of the Central Intelligence Agency, which highlights the fact that we are overdue for a new Ice Age. Many climatologists believe that since the 1960s, the world has been slipping

towards a new Ice Age ... the evidence suggests that the change will be a return to a climate that was dominant from the seventeenth Century to about 1850. Soviet weatherman Mikhail Budyko believes that a 2.8°F drop in the average global temperature would start glaciers on the march. If the temperature should fall by another 0.7°F, it could usher in a ninety-thousand year tyranny of ice and snow. (The weather conspiracy: the coming of a new Ice Age. The Impact Team, 1977).

Messages from human history

The NASA satellites recording temperatures in the Earth's atmosphere have clearly documented a pause in average global atmospheric temperatures between 1998 and 2017. There are still variations from year to year, and local anomalies, but the average trend is sideways. If any year in the past 19 years has been higher than another, it is only by fractions of a degree. This trend has not only been reported by NASA, but also by the British Hadley Centre.

The modern media, when some recent temperature record is broken, make statements to the effect: *"We are in uncharted waters!"* or *"This is the a sign of a climate Armageddon".* But, these temperatures need to be placed within the context of the climate during the past 10,000 years, since the retreat of mid-latitude ice sheets. *When this is done it is clear we are not in unprecedented territory in terms of temperature change.*

The period between 8,000 and 4500 years ago was actually far warmer than today and is called the Holocene Climate Optimum or the Holocene Thermal Maximum. Evidence for this is indisputable. Sea level was over 2 metres higher than today. Pine trees grew in Scotland 650 metres above the stunted trees of today. Mistletoe and sub-tropical plants grew in Southern Sweden. Oak, elm and spruce were the common trees in northern European forests. Even though there have been warm periods since that time, the general temperature trend has been downwards. However, the Minoan Warm Period 4000 years ago, and the Roman Warm period 2500 years ago were warmer than the Medieval Warm Period, 1000 years ago.

During the Minoan Warm Period millet, which now grows in France, flourished in southern Sweden. Then in the Roman Warm Period, 2000 years ago, vineyards were in southern England (Sussex and Kent) and middle England (Northampton) and even as far north as Hadrian's Wall near the Scottish border. At the same time, England was an important granary for the Western Roman Empire (Plimer, 2009).

Climate comparisons between the Northern Hemisphere and Southern Hemisphere at equivalent latitudes during any of these warm periods can easily overlook obvious differences with regard to land distribution (68% versus 32%), and in the mixing of polar and mid-latitude oceanic currents. These significant differences imply that the recovery of each hemisphere towards warmer conditions after an ice age or cold period may be slightly different in time and effect.

There is considerable historical data enabling comparisons between the climate in medieval times and that of today. The evidence for the Medieval Warm Period in the Northern Hemisphere is strong and lies in areas between 50° north (for example, southern England) and 65° north (for example, southern Greenland, southern Norway and Sweden). Apart from human historical records there is considerable land evidence of flora and fauna changes in this warm period. The Vikings diaries describe diary farms in Greenland; in Estonia detailed harbour records show that ports were closed by ice for shorter periods; and in Japan the first Spring cherry blossom dates recorded by monks show that Spring arrived earlier.

In China, there are over 50 peer reviewed scientific articles on the Medieval Warm Period describing sediment analysis, tree ring data, chemical analysis of limestone formation in caves (for example, the Buddha Cave or the Jingdong Cave), the shift of sub-tropical plants to northern China, and the extent of glacial retreat in north-west China. Chinese authors report warming rates in the 12th and 13th centuries greater than those of the 20th century and their data show that the 20th century temperatures are not abnormal:

> ...the temperature of the 20th century in Eastern China is still within the variability of the past 2000 years (Ge et al., 2010).

It has taken some time to find evidence of the Medieval Warm Period in the Southern Hemisphere. Logically, the evidence should also lie between latitudes 50° south and 65° south. However, in the Southern Hemisphere these latitudes lie in areas covered by the Southern Ocean. Here there are no significant landmasses only small sub-Antarctic Islands and there is no recorded human history. In this respect the difference between the hemispheres becomes obvious when Britain is placed at its equivalent latitude in the Southern Ocean (Diagram 4.3). So, some arguments about the lack of historical evidence for the Medieval Warm Period in the Southern Hemisphere are based on unfair comparisons of latitudes where there are significant differences in empirical land evidence.

Diagram 4.3: A map placing The British Isles in their equivalent position in the Southern Hemisphere (adapted from Google Earth).

Evidence for the Medieval Warm Period in the Northern Hemisphere is found mainly between 50° N and 65°N. The equivalent zone in the Southern Hemisphere is south of South Africa, Australia and South America. The main evidence for warming in the Southern Hemisphere during the Medieval Warm Period comes from fossils buried on the ocean floor where isotope analysis shows a warmer deeper ocean at that time.

Despite this bias there is some evidence in the Southern Hemisphere north of 50°S for a warmer period in the Medieval Warm Period such as historic cave temperature data from South Africa and tree ring and glacial evidence from Patagonia and Chile. Professor Peter Tyson from the University of Witwatersrand, South Africa, has analysed the past temperature history of caves in South Africa and these reflect the Medieval Warm Period (Tyson et al., 2000). Dr Ricardo Villalba from the University of Colorado has also shown a Medieval Warm Period from analysis of tree ring data in South America (Villalba, 1994).

However, the key evidence for a Medieval Warm Period in the Southern Hemisphere does exist in the same comparative latitudes; it is just not on land. The evidence comes from the analysis of deep ocean sediments. Different forms of oxygen (isotopes) are used by scientists as a clever thermometer to indicate the temperature of the water when the shells of certain fossils were formed. Professor Yair Rosenthal from Columbia University, New York, and other scientists have analysed different types (isotopes) of oxygen in microfossil shells from sediments collected from the Pacific Ocean floor.

The data show that Deep Water masses in both the North Pacific Ocean (Northern Hemisphere) and the South Pacific Ocean (Southern Hemisphere) were 0.9°C warmer during the Medieval Warm Period than during the following Little Ice Age, and even 0.65°C warmer than today. This Deep Water must come from the Antarctic region because as pack ice freezes the ice is less salty and the remaining heavier, saltier water sinks and gradually covers the ocean floor all the way to the Tropics. So, at anytime when this Deep Water is warmer than in other periods, the Southern Ocean near Antarctica, that is the source of this water, also has to be warmer.

Such evidence clearly dispels arguments that some of these past warm periods, such as the Medieval Warm Period, were simply Northern Hemisphere events. In addition, the data clearly indicate not only a Medieval Warm period in the Southern Hemisphere, but also that the Deep Ocean Heat Content at the present time is neither unprecedented nor abnormal. Indeed, the Deep Ocean Heat Content in both hemispheres today is not only less than in the Medieval Warm Period, but it is well below the Deep Ocean Heat Content in the very warm period scientists call the Middle Holocene Thermal Maximum, which occurred 10,000 years ago (Rosenthal et al., 2013).

Messages from glaciers

European alpine glaciers provide an excellent record of glacial retreat during warm periods over the past 4000 years. Such retreat enables us to compare present temperatures with temperatures in the past. The history of the glaciers in the French and Swiss Alps clearly shows that *modern glacial retreat is not an unprecedented event*. Geologists, such as the famous Professor Hanspeter Holzhauser of the University of Bern, Switzerland and Dr Melaine Le Roy from the University of Savois in Mont Blanc, France have measured the advances and retreats of Alpine glaciers (Diagram 4.4).

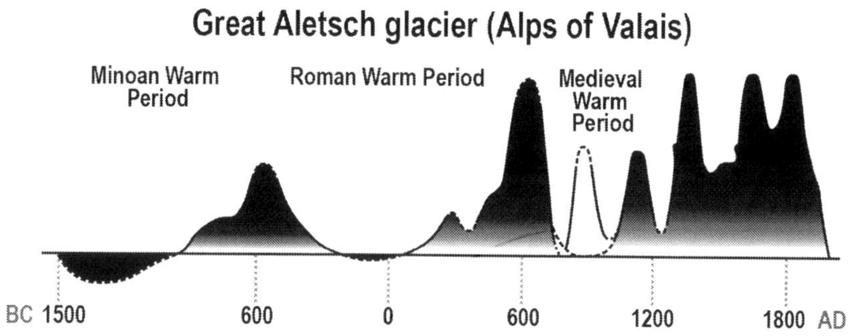

Great Aletsch glacier (Alps of Valais)

Diagram 4.4: The retreats and advances of The Great Aletsch Glacier over the past 3500 years as estimated by Professor Hanspeter Holzhauser of the University of Bern, Switzerland.

This graph show significant warm phases in the Swiss Alps centred around 1500 BC, 100 BC, and 900 AD. The lengths of past warm periods vary considerably between 300 and 600 years. However, it is clear from this diagram that the extent of modern glacial retreat is not unprecedented. The diagram clearly shows that the Roman Warm Period was longer than the Medieval Warm Period. There are more diagrams showing the past history of other Swiss glaciers by Professor Hanspeter Holzhauser (1997). There are also similar diagrams by Dr Melanie Le Roy (2015) showing the history of glaciers in south eastern France such as the Argentière, Bossons and Mer De Glace glaciers.

The European Alps were nearly glacier free in the Minoan Warm Period, the Roman Warm Period and also the Medieval Warm Period. The large Roman bridges in Syria over dried river beds such as at Uthma, Maharda or Djemarin are clear evidence of a wetter climate in the Middle East during the Roman Period. The climate was much warmer than today and caused the glaciers in the European Alps to retreat. As those glaciers retreated they exposed logs from even earlier times when these regions were forested and glacier free. These exposed logs are telling us that nature will repeat itself, and that more of those Alps may be forested again if the present warming continues. One wonders what vegetation Hannibal would have described when he crossed the Alps and invaded Italy in 218 BC with elephants and 90,000 men.

In the modern warm period, glacial retreat started in the 18th century, well before any rise in carbon dioxide levels. In the 19th century it was common knowledge that glaciers were retreating worldwide, and that the climate was changing. This news was making its way even to small provincial newspapers. For example, Braidwood was then a small, remote gold mining town, 200km south of Sydney, Australia, with a population around 1500, and even the Braidwood Dispatch and Mining Journal of September 28, 1910 reported:

> *Except over a small area, it is understood the glaciers of the world are retreating to the mountains. The glacier on Mount Sermiento in South America, which descended to the sea when Darwin found it in 1836, is now separated from the shore by vigorous timber growth. The Jacobshaven Glacier in Greenland has retreated 4 miles since 1850, and the East Glacier in Spitzbergen is more than a mile away from its original terminal moraine. In Scandinavia the snowline is further up the mountains and the glaciers have withdrawn 3000 feet from the lowlands in a century. The Arapoe Glacier in the Rocky Mountains, with characteristic American enterprise, has been melting at an alarming rate for several years.*

Up the wrong tree

The temperature rise over the past 150 years was made to look unprecedented by Professor Michael Mann of Pennsylvania State University. He downgraded the temperatures in the Medieval Warm Period, and compared the temperatures during the past 1000 years to a hockey stick. The curved blade at the base showed the steepness of present temperature rise and the straight stick represented the steady temperatures of the past 1000 years. Mann argued, mainly from his analysis of tree-rings in high altitude bristlecone pines in the south-western United States, that the so-called Medieval Warm Period (900 AD to 1200 AD) was just an isolated local warming event in the Northern Hemisphere, and that there was no general worldwide global warming at that time (Mann et al, 1999).

There is clear evidence of warming during the Medieval Period in both hemispheres; Michael Mann's argument is dead in the water. However, the Third Assessment Report of the IPCC in 2001 used Mann's ideas to show that the Medieval Warm Period was an isolated European event and that the average worldwide temperatures had been flat for thousands of years. *Such a false view was used to highlight the unusual warm trends in the 20th century and the strong possibility that modern warming was due to the increased greenhouse gases.* It should be noted that Professor Hans Von Storch, Professor of Meteorology of the University of Hamburg, was a lead author in the Third Assessment Report (2001), and *strongly opposed any use of Michael Mann's hockey stick description* of world temperatures *in the past 2000 years.* He was overruled, and, as a result, he was not invited to participate in the 2004 IPCC Assessment Report.

Final reflections

Modern-day temperatures are well within the temperature bands of warming periods during the past 4000 years, and cannot be described as *unprecedented.* Why then the panic? Why the misinformation? And why just argue about the Medieval Warm period when there were other warm periods in the past few thousand years that were even warmer than today?

When we place the temperature trends in the past 150 years in perspective:

- There has been a warming trend for the past 300 years.
- The data for the past 150 years show a stepwise warming temperature pattern.
- The data for the past 150 years show there has been 70 years when global temperatures paused, and yet carbon dioxide levels continued to rise throughout this whole period.
- The data for the past 150 years show the temperature gradients in the three warming periods have been similar, without any acceleration.
- Present temperatures are similar to those in the Medieval Warm period (1000 years ago), and not as high as the Roman Warm Period (2000 years ago).
- The history of glacial retreat in the European Alps is well documented, and shows the present glacial retreat is not unprecedented.

The world may continue this warming trend, carbon dioxide levels may continue to rise, glaciers may continue to retreat, more temperature records may be broken, but nature is more like a person going up and down a staircase. *Nature does not have its foot on a temperature accelerator.*

Chapter 5.

A weather Armageddon

The weather scare-line

When extreme weather events occur, the media link severe weather, described as unprecedented or exceptional, to increasing carbon dioxide levels.

The narrative first unfolds when some climate scientist makes dramatic statements on the reason for recent weather and the media publicise the story. Understandably, the media will enhance the story to sell more papers, but, even more so if a media group is aligned to various conservationist groups. After that, various lobbyists take up the storyline. For example, any talk of unheard-of weather linked to increasing greenhouse gases strengthens the political clout of those wanting to stop the use of fossil fuels.

In all this talk, the lines between the media, various interest groups, and the body politic are very blurred and full of misleading general statements.

There is a simplistic view that higher temperatures mean more evaporation of surface waters, more water in the atmosphere and therefore more rain and storms. It is not as simple as that because a rise in the relative humidity in the air slows down surface evaporation. Relative humidity describes how much water is in the air at a certain temperature and pressure compared to the maximum water vapour the air could hold under those conditions. That is why relative humidity is described as a percentage or fraction. Professor Richard Lindzen from the Massachusetts Institute of Technology, who wrote 200 papers on atmospheric physics before retiring, pointed out that a small increase in relative humidity from 80% to 83% is enough to eliminate any increase in evaporation rates should the air temperature increase 3°C (Lindzen in Moran 2014).

Examples of statements reflecting simplistic views about the effects of increased evaporation and a warmer atmosphere on storm frequency include:

However, because a warmer atmosphere holds more water vapour, global warming must also increase the intensity of the other extremes of the hydrologic cycle – meaning heavier rains, more extreme floods, and more intense storms driven by latent heat, including thunderstorms, tornadoes, and tropical storms (James Hansen, 2009. *Storms of My Grandchildren*).

Climate change and population growth will hugely increase the risk to people from extreme weather, a report says. The Royal Society warns that the risk of heat waves to an ageing population will rise about ten-fold by 2090 if greenhouse gases continue to rise. They estimate the risk to individuals from floods will rise more than four-fold and the drought risk will treble (Roger Harrabin–BBC environmental analyst November 27, 2014).

A climate that increases in temperature will mean more extreme and frequent storms, more flooding, rising seas that submerge Pacific islands … The incredible natural glory of the Great Barrier Reef is threatened (Barack Obama, Talk to university students, November 16, 2014, Brisbane, Australia).

The above statements look very logical, yet are incorrect because *there has been a 40% increase in carbon dioxide levels over the past 150 years without any meaningful increase in storm frequency or intensity.* Many people have not looked carefully at long-term weather records and just presumed they knew the answer. Have we let our imagination run away with itself? Have we been panicked by the storyline of increasing apocalyptic weather? Are we like Chicken Little, Henny Penny, Ducky Lucky and Goosey Loosey who believed the sky was falling when a little acorn fell on Chicken Little's head? They were sucked in by panic, then took the protection offered by the fox in his lair, and were eaten.

Destructive storms

In a large nonlinear climate system if one cherry-picks singular events, one will find record-type events, but the occurrence of such events does not necessarily mean they are occurring with increasing frequency. That will only be true after complete data sets are examined. When this is done it is clear that *the increasing storm frequency narrative is based on cherry-picking data.*

European historical records indicate that there has been very severe weather in recent cold periods such as the Little Ice Age between the 14th and 16th Centuries AD. Unfortunately, there is not enough data to show the frequency of such events. Professor Hubert Lamb, who founded the Climatic Research Unit at the University of East Anglia, England, in his book *Historic Storms of the North Sea, British Isles and Northwest Europe* comes to the conclusion that various storms in the Little Ice Age were much more severe than recent storms (Lamb, 1991).

Edward Bryant, an Associate Dean of Science at Wollongong University in Australia, in his book *Natural Hazards* describes many severe weather events between the 13th and 18th Centuries (Bryant, 2004):

- Four storms along the Dutch and German coasts in the 13th century killed at least 100,000 each. The worst is estimated to have killed 300,000.
- Much of the coastline of northern Europe owes its origin to this period of storms. For instance, storms reduced the size of the island of Heligoland from 60 sq km to 1 sq km. Similarly the Great Drowning Disaster of 1362 eroded 15 km landward of the Danish coast, destroying over 60 parishes.
- North Sea storms in 1099, 1421 and 1446 killed 100,000 each in England and in the Netherlands.
- By far the worst storm was the All Saints Day flood of 1570, when 400,000 people were killed throughout Western Europe.
- Other storms with high death tolls occurred in 1634, 1671, 1682, 1686, 1694 and 1717.
- The Great Storm of 1703 sank virtually all ships in the English Channel, with the loss of 8000 to 10,000 lives.

In 1694, if we were in the Kinnard clan, what would we have thought was happening with the climate when a violent sandstorm lasted two days and buried 20 sq km of our fertile Scottish farmland, not with one or two metres, but *with 30 metres of sand* (Doe, 2006).

There are many records of storms in European manuscripts, but there are also very detailed records from China. The largest typhoons (*jufeng*) are recorded in the official histories of the Chinese Dynasties, but more detailed jufeng records are in the Diaries of the Emperors *(Veritable Records)* and in the astonishingly accurate records of the semi-official local government gazettes *(fang zhi)*. For example, Professor Kam-Biu Liu, of Louisiana State University and his colleagues show that the Guangdong gazettes alone record 571 typhoons in a 935-year period from 975 AD to 1909 AD. During that time, the most active typhoon periods were 1660–1680 and 1850–1880 (Kam-biu Liu at al., 2001).

It is only in the past 200 years that there have been careful recordings of weather all over the world. Those data show no increased frequency of severe storms, but because of the exponential increase in human populations and associated infrastructure (especially in coastal areas), there has been a huge increase in *storm damage* and *loss of life*. But, how can an argument jump from *increasing damage, rising financial costs, higher loss of life* to *more severe storm frequency*? That is lazy logic!

Despite this lack of logic, the idea of more frequent and more violent weather has spread like a virus and has infected the minds of intelligent people everywhere. It is misleading to say:

> The number of extreme weather events seems to have quintupled since the 1950s, according to the insurance company Munich Re (Monbiot, 2006).

But, what do the data say?

US hurricanes

The National Oceanic and Atmospheric Administration of the United States (NOAA) records hurricane frequency in the eastern United States. The data for 130 years shows no upward trend in major storms. There is an upward trend in short two-day storms but NOAA reports that is due to better observational systems. As of January, 2017, there had been 3,365 days (9.2 years) since a Category 5 hurricane hit the eastern US mainland! This is by far the longest such stretch since record-keeping began in 1900, if not since the American Civil War (Diagram 5.1.)

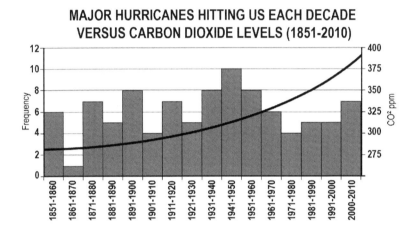

Diagram 5.1: NOAA Graph of Atlantic storms each 10 years since 1880.

NOAA reports that the frequency of short based storms has increased due to better reporting. With regard to major storms NOAA reports: "We find that, after adjusting for such an estimated number of missing storms, there is a small nominally positive upward trend in tropical storm occurrence between 1878 and 2006. But statistical tests reveal that this trend is so small, relative to the variability in the series, that it is not significantly distinguishable from zero. Thus the historical tropical storm count record does not provide compelling evidence for a greenhouse warming induced long-term increase." (Vecchi and Knutson, 2015).

Australian cyclones

The Australian Bureau of Meteorology has records of tropical cyclones with pressures less than 970 hectopascals (hPa). There is no general upward trend in cyclone frequency in the past 42 years (Diagram 5.2.). The cyclones either strike the west coast of Australia or sweep across northern Australia into the Coral Sea where they often back-track and wreak havoc along the northern coast of eastern Australia.

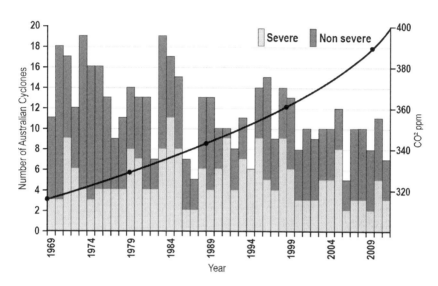

Diagram 5.2: The frequency of tropical cyclones in Australia between 1969 and 2012 as recorded by the Australian Bureau of Meteorology.

The Australian cyclones in this graph are those with severe low pressures less than 970 hectopascals. It is clear that there is no increasing trend towards more severe tropical cyclones during the past 42 years; if anything there has been a declining trend. There is no relationship to the increasing trend in carbon dioxide levels that rose 19% from 324 parts per million to 400 parts per million during this period.

The Wind Intensity Index

The National Oceanic and Atmospheric Administration of the United States (NOAA) records the highest winds in cyclones and hurricanes every six hours. In the Tropics. This is known as the Accumulated Wind Intensity Index. Records kept over the past 42 years show no rising trend in wind intensity even though carbon dioxide levels increased throughout this period (Diagram 5.3).

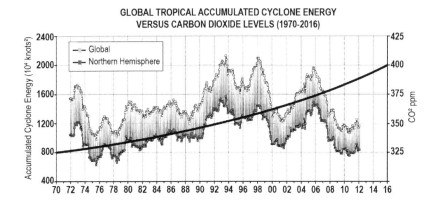

Diagram 5.3: NOAA Graph of the Global Tropical Accumulated Cyclone Energy Index 1972–2014.

The National Oceanic and Atmospheric Administration of the US(NOAA) records the highest winds every 6 hours in tropical cyclones worldwide and in the Northern Hemisphere. There is no global pattern of increased cyclone wind intensity during this period. There is obviously no correlation with carbon dioxide levels, as carbon dioxide levels increased by nearly 20% during this period.

US Tornadoes

Sixty years of strong to violent US tornado data kept by the National Oceanic and Atmospheric Administration of the United States show no upward trend in frequency or severity (Diagram 5.4). Note that very small tornadoes are not in the graph because better weather radar is detecting them and they were not well recorded years ago.

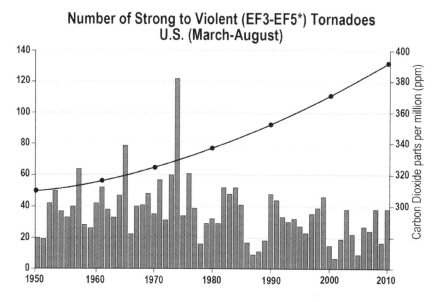

Number of Strong to Violent (EF3-EF5*) Tornadoes U.S. (March-August)

Diagram 5.4: NOAA Graph of US Tornado frequency Force 3 and above 1950–2010.

This graph shows the annual frequency of serious tornadoes in the US over 60 years (1950–2010). Tornadoes are defined according to the damage they cause in a scale called the Fujita Scale. Their damage is related to the width of the tornado and its speed. F-3 tornadoes have wind speeds of 254–332 km/h, whereas F-5 tornadoes have wind speeds of 419–512 km/h. There is clearly no relationship between violent tornado frequency and rising carbon dioxide levels in this 60-year record.

The parched Earth

Droughts are also classified as extreme weather events. Indeed, droughts have destroyed civilisations in the past. For example, fossil evidence indicates a 200-year drought in Sierra Nevada between the 9th and the 12th Centuries AD. Mega-droughts are thought to be responsible for the collapse of several empires, such as the Mayas in Middle America in the 9th century AD, the Yuan Dynasty in China and the Khmer Empire In Cambodia in the 14th century AD (Fagan, 2004).

The US National Climate Data Centre has a record of droughts in the US over the past 112 years (Diagram 5.5). There is no historical evidence in this period of an increase in droughts and no recent mega-drought (a drought lasting more than 20 years).

Recently, Ben Cook, a climate scientist at NASA's Goddard Institute for Space Studies and the Lamont-Doherty Earth Observatory at Columbia University in New York City, forecasted a mega-drought late in this century, but this forecast was based on models with carbon dioxide levels tripling within the next 90 years (Cook et al., 2015). Ben Cook remarked:

> Natural droughts like the 1930s Dust Bowl and the current drought in the Southwest have historically lasted maybe a decade or a little less. What these results are saying is we're going to get a drought similar to those events, but it is probably going to last at least 30 to 35 years.

Greenhouse gases have risen ≈40% in the past 150 years, yet Cook has confronted the public with a scenario based on a prediction that the Earth's carbon dioxide levels will triple within the next 90 years. This catastrophic story-line is an example of fairy-tale science; it does nothing for the credibility of NASA.

No trend in the amounts of extremely dry periods in USA
Percent of US Severely to Extreme Dry
Source: National Climate Data Centre

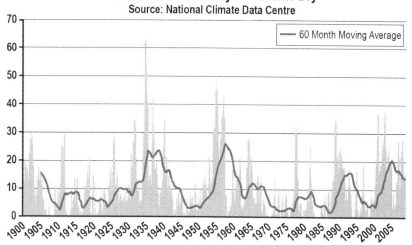

Diagram 5.5: US Drought frequencies from the US National Climate Data Centre between 1900 and 2010.

This diagram from the US National Climate Data Centre shows the frequency of US droughts in the 20th century. There is clearly no relationship between US droughts and the rising carbon dioxide levels in the past 110 years. The severity of the drought in the 1930s stands out. During that drought Texas, Oklahoma and parts of Kansas, New Mexico and Colorado were reduced to dust bowls. The drought was worsened by poor farming methods and its effects are dramatically portrayed in John Steinbeck's famous novels, Of Mice and Men (1937) and Grapes of Wrath (1939). These novels examined the lives of migrant workers and the farming communities displaced by this great drought.

The scorched Earth

Bushfires are also classified as extreme weather events. One would expect bushfire frequency to increase with global warming and that areas prone to bushfire would change, even increase, as climatic zones shift. Such changes are not proof that the main reason for modern global warming is increased greenhouse gases; that is a separate argument because bushfire frequency could increase whatever the reason for global warming.

Bushfires have been common throughout history. There are also extensive records of bushfires being used for land management by the Aboriginal peoples in Australia (fire-stick farming).

Early European settlers recorded huge fires in the grasslands (prairies) of Canada and the US. As settlers moved westward in the 1700s, the wanton killing of bison that grazed on these prairies increased the risk of fire from human intervention and lightning. The worst bushfire in the United States was in October, 1871 when more than 1500 people died in the Peshtigo forest fire. The firestorm threw railway carriages and houses into the air. On that day there were fires in the forests around Lake Michigan and Lake Huron and the famous Great Chicago Fire killed 300 people and left 100,000 people homeless.

Some recent dramatic bushfires were:

- Indonesia 1997–1998: These fires lasted for 8 months and were probably the largest in recorded history. The haze was clearly visible from space and millions of people throughout South-East Asia suffered from smoke pollution as 80,000 sq km (20 million acres) of rainforest were destroyed.

- Australia 2009: Australia is very vulnerable to bushfires and the risk is intensified by the flammable oil in the leaves of its eucalyptus trees. During the Victorian fires of 2009, 173 people lost their lives. Most famous were the Black Friday fires of 1939 in Victoria and South Australia that burnt out 30,000 sq km (seven million acres) of forest and killed 73 people.

- Greece 2009: In the summer of 2009 huge forest fires broke out 40 km north of Athens in Greece. Over 210 sq km (52,000 acres) of forest were destroyed and 14 small towns were destroyed.

- United States 2012: Between May and July 2012 a series of fires engulfed sections of Colorado. Huge areas of forest were destroyed: Little Sand–57 sq km (14,285 acres); High Park–353 sq km (87,250 acres); Last Chance–182 sq km (45,000 acres); Fern Lake–14 sq km (3,500 acres); Pine Ridge–53 sq km (13,920 acres); Weber–40 sq km (10,000 acres). In the Waldo Canyon Fire 32,000 residents of Colorado Springs were evacuated and 350 homes destroyed.

- Russia 2015: Forest fires in the Chita region of Eastern Siberia in April 2015 killed 29 people, wounded over 460 people and left 6000 people homeless.

Unfortunately, some recent loss of life in forest fires has been due to population increase. There are now more urban fringe dwellers living close to cities in bushy areas without adequate bushfire clearance zones. Indeed, in some areas, vegetation legislation can make it impossible for individuals to adequately clear around their woodland homes, and sensibly protect themselves against bushfires.

An ever restless climate

Shifts of weather patterns are to be expected as climate changes. Even though storm frequency may increase in certain places and decrease in others, as climatic zones shift, frequency graphs show no overall trend upwards. A famous example of climatic shift was the desiccation (drying out) of the (now) Sahara Desert area after the ice sheets retreated from Europe. At that time, the belts of rain, that were frontal systems in front of that ice sheet and which had turned the Sahara desert into a lush grassland, moved northwards. A modern example would be lower rainfall in south-east and south-west Australia. Upon any contraction of the permanent cold high-pressure system over the polar Antarctic landmass, the low pressure systems in the Southern Ocean retreat southwards. This means lower rainfall in southern Australia.

It can be argued that there has been no increase in severe weather events in recent times because the global warming was less than 1.0°C in the past 150 years and that it takes higher rises in temperature to shift severe weather patterns.

It is possible that if the world continues to warm higher temperatures in tropical regions may increase the intensity of tropical cyclones (when they occur), but, at the same time, there may be less severe weather events and lower wind-speeds in mid-latitude areas as the temperature gradient towards the poles decreases; that is, a lower equator to pole temperature gradient (EPTG). This would be due to lower pressure differences between the equator and the mid-latitudes and consequently less barometric instability. It is also possible if the world cools that lower temperatures in tropical regions may decrease the intensity of tropical cyclones, and lead to a more severe EPTG as more polar weather intrudes into the mid-latitudes and brings with it an increase of severe cold weather.

These arguments seem logical, nevertheless we are still dealing with a nonlinear system where complex relationships between many factors leads to unexpected outcomes. *Arguments that severe weather events should increase in severity and frequency this century due to global warming are false, and statements that the increases could even be ten-fold are plainly fairy-tale science.* After all, the world has seen a 40% increase in carbon dioxide levels in the past 150 years with no noticeable effect on the frequency of severe weather events.

The fictitious storyline

Data show no recent change in the frequency of cyclones, tornadoes, hurricanes and droughts, and yet carbon dioxide levels have increased over 40% in the past 150 years. As there has been no link between rising levels of greenhouse gases and the increased frequency of severe weather during this period, why should there be a strong causal relationship between them in the next 100 years?

We have been told that the incidence of severe weather events has actually increased in frequency when it has not. The media, politicians, and various scientific bodies have misled us. We have no alternative, but to take any statements about the increase of severe weather in the future with a pinch of salt.

The most extraordinary aspect of the recent climate debate is that many people who have made misleading statements about the frequency of severe weather events are intelligent, even famous people who have become gullible robotic lemmings; just repeating the statements of others. In addition, there are other spokespeople willing to promote false stories to bolster their agenda. Furthermore, in the middle of this maze of argument, critical media analysis, based on solid scientific data, is the exception, rather than the rule.

Chapter 6.

Sea level rise – steady as she goes

Aquaphobia

Whenever the world warms, high altitude glaciers retreat and seawater expands slightly. There may be some melting on the Greenland Ice Sheet and the speed of its coastal glaciers may rise. In Antarctica surface temperatures are too low for any surface melting on the ice sheet interior but the collapse of large coastal ice shelves may mean that ice streams from the interior that are moving due to natural plastic flow may have an easier and slightly faster pathway to the oceans. Most of the panic in the climate debate has been about rising sea levels and the effects this will have on human development. Projections are made about possible sea level rise. Newspapers print photos showing future sea level lines on famous surf beaches; like those in California, or along the French Riviera, or Copacabana Beach in Rio De Janeiro, or Bondi Beach in Sydney, Australia. This propaganda strikes the nerve of public sentiment.

However, how is sea level measured? For many years seaside ports kept their own records of sea level and some of these tide gauge records go back over 200 years. *Using satellites to measure world sea level only began 25 years ago.* All sides of the climate debate agree that the average sea level rise in the 20th century was around 15–17cm (that is between 1.5mm/yr and 1.7mm/yr). This figure is calculated by averaging tide gauge data around the world.

Despite this, it is difficult to estimate where this extra water came from. We know the ocean expands as it warms, and we know sea level rises as glaciers in say Europe, the Himalayas, or the Andes melt, and we also know sea level rises if ice sheets in Greenland or Antarctica melt. However, we do not know what proportion each has contributed to sea level rise in the 20th century.

It is not easy to measure sea level at any particular time. The ocean is not flat like still water in a bath tub. As the Moon travels around the

Earth, its gravity tugs on the ocean, raising its height. There are two high tides each day. One occurs when the Moon is overhead, and the other, interestingly, when the Moon is on the opposite side of the Earth. Tidal variations normally range from 1–2 metres, the highest around 12 metres and the lowest less than one metre.

But, as the water bulges due to the Moon's pull, there are many other factors that influence the final result. These include: long amplitude ocean waves; complex longitudinal currents along coastlines; tidal forces amplified by the spinning of the Earth; and persistent winds that can pile up water over large distances. *So there is no such thing as a uniform world sea level.*

No pedal on the metal

The debate about sea level rise during the past 150 years mirrors the debate about temperature increase in that period. In both debates, there are discrepancies between future sea level predictions derived from projecting forward historical records and sea level rise forecasts made by computer models. For example, scientists looking at historical tide gauge data do not see a long-term accelerating trend, while the scientists using computer modelling and satellite data predict sea level acceleration.

The IPCC's precautionary scenarios published since the 1990s predict sea level to rise between 26 cm and 82 cm by the year 2100; almost 2 to 5 times current rates between 15 and 17 cm/100 years. These higher scenarios can only occur if sea level rise accelerates.

Scientific articles examining tide gauge data find little if any acceleration in sea level rise in the past 160 years. Even a frequently quoted article by Dr John Church and Dr Neil White, from the Australian Commonwealth Industrial Research Organisation (CSIRO) finds an acceleration so small that it is open to debate as to its timeframe and statistical analysis (Church and White, 2006).

In 2007, Simon Holgate's examination of nine worldwide tide gauges over 100 years showed a steady sea level rise of 2.03 mm/year between 1904 and 1953 (20 cm/100 years) and then a slight deceleration to 1.45 mm/year between 1954 and 2003 (14.5 cm/100 years). Professor John Houston of the world famous US Army Corps of Engineers, and Professor Robert Dean of the University of Florida, have released an analysis of long-term tide gauge data from coasts of the US, Western Europe, Australia and New Zealand (Houston and Dean 2011, 2013). Their examinations of tide gauge data over the past 160 years indicate that sea level has changed through various short periods of acceleration and deceleration, but there has been no long-term acceleration during that period.

Professor Bob Carter, formerly of James Cook University, Queensland, and who served on the Australian Research Council, summarised scientific reports on sea level rise in his book and clearly shows there is no solid evidence for sea level acceleration (Carter, 2010, pp.92–96). There is also another scientific study by Phil Watson, a senior Australian coastal engineer working for the New South Wales Government. He supports the analysis of professors Houston and Dean, as his analysis of old reliable Australian and New Zealand tide gauges indicates a recent, weak deceleration of sea level, at least in this region during the past 60 years (Watson, 2011).

Even while the historical records show no acceleration of sea level, scientists and consultants appear to accept the IPCC predictions of high sea level rise when they develop urban planning policies. In an effort to prove *conservative* some environmental engineers forecast even *higher* sea level rises of between two and five metres to the year 2100.

Environmental engineers, who are often not that familiar with the science of climate, look at the IPCC position and include sea level acceleration in their planning reports; some are clearly confused. For example, this rather perplexing statement appears in a report to a local Australian regional council. Doug Lord and David Wainwright believe something is happening that seems from the data not to be happening; an argument that defies normal logic:

> *Although the inclusion of results from many models generates uncertainty, the overall projection of an accelerating future sea-level rise is clear, even if that acceleration cannot yet be unequivocally proven based on the presently available measured record* (Lord and Wainwright, 2014).

Satellites or tide gauges?

In order to obtain global coverage, measurement of sea level rise using satellites commenced 25 years ago. This project has significant problems. The University of Colorado publishes altimeter recordings of sea level rise, derived from a series of satellites, replacing each other over time. The current satellite is Jason 3. In satellite altimetry, all measurements are made in reference to a mathematical shape, called a geoid, that represents the surface of the Earth, and known as a Terrestrial Reference

Frame (TRF). Any error in calculating this TRF will corrupt the data. Using this method, the mean rate of sea level rise between 1992 and 2013 was 3.2 mm/year (32 cm/100 years) and then the rate slowed a little in the decade ending in 2013 (Xianyao Chen et al., 2013).

However, the data, as calculated by the present satellite Jason 3, has diverged so far from long-term and reliable tide gauges all around the world, it is evident there are serious problems with the instrumentation and computer modelling associated with the present satellite altimetry systems. Professors Houston and Dean echo their concern about this satellite altimetry:

> ... altimeter and tide gauge measurements were in good agreement up until 1999 and then began to diverge with the altimeters recording a significantly higher sea-level trend than worldwide tide gauge records (Houston and Dean, 2011).

In July 2014, Klaus-Eckart Puls of the European Institute for Climate and Energy reported on its website that:

> ... Moreover, there are indications that the satellite data (showing a higher [double] rate of increase) are significantly over-corrected.

There is another means of measuring sea level, and that is by weighing the ocean. In 2002 a joint mission between NASA and the German Aerospace Centre launched twin satellites to measure the Earth's gravity. The mission is known as GRACE–the Gravity Recovery and Climate Experiment. The GRACE satellite measurements indicate sea level rise of 1.6 mm/year (16 cm/100 years), and are close to the average tide gauge sea level rise measurements of 1.7 mm/year (17 cm/100 years).

NASA publicly admitted in 2012 there are serious problems with the present satellite system when it made its financial proposals to the US Congress for a new satellite system called GRASP:

> Thus, we assess that the current state of the art reference frame errors are at roughly the mm/yr level (one mm/yr is an error of 10 cm/100 years, and so forth) making observation of global signals of this size very difficult to detect and interpret. This level of error contaminates climatological data records, such as measurements of sea level height from altimetry missions, and was appropriately recognised as a limiting error source by the NRC Decadal Report and by the Global Geodetic Observing System (Yoaz Bar-Server et al., 2012. Jet Propulsion Lab. Caltech under contract to NASA).

Dr Yoaz Bar-Server, at the NASA Jet Propulsion Laboratory in California, is leading the planning for the new GRASP satellite to replace Jason 2. In September 2014, he emailed to this author:

The discrepancy between long-term tide gauge records and ocean altimetry records is indeed a primary motivation for GRASP. I hope we'll have good news in about a year from now.

The inherent problems with the satellite data highlight the importance of the long-term tide gauge data at regional sites like Fort Denison in Sydney Harbour, Australia. *This is recognised as one of the world's best kept tide gauges, supplying primary data.* This tide gauge is also on a stable land mass that has not been subject to recent movements in the Earth's crust.

Despite problems with the sea level satellite data, some scientists use the satellite sea level rise data as a benchmark and then adjust tide gauge data upwards to the satellite data. But once that tide gauge data is averaged with or merged with satellite data it is being totally corrupted. Since the average Fort Denison sea level rise rate, as measured by the tide gauge over 100 years, is less than 10 cm/100 years, it is *absurd* to adjust that data *upwards* to the satellite data from the nearby Tasman Sea. It is *equally absurd to average* the Fort Denison tide gauge data (that are less than 10 cm/100 years) with the sea level rise estimates from the satellite data that are around 30 cm/100 years.

Coastal authorities should use long-term regional tide gauges, as the primary source for their development policies along beaches and headlands. Satellite altimetry, further refined by GRASP, promises an order of magnitude improvement over present techniques. That data, still some years away, will complement, but not replace, reliable regional tide gauge records.

There are now special tide gauges (SeaFrame gauges) in Australia and the mid-Pacific islands attached to fixed sites that transmit tidal data continuously via satellite to a central measuring station. For example, the SeaFrame tide gauge data at Port Kembla (one of the three SeaFrame sites on the east coast of Australia) are transmitted to an Australian research organisation, and then merged with other Australian SeaFrame data, and then with the satellite data. As published, the merged data show a sea level rise in line with the Jason satellite data and reflects the same errors as the Jason satellite series. For example, the Port Kembla SeaFrame shows a sea level rise three times more than the tide gauge data at reliable Fort Denison in Sydney (100 km to the north) or at the Australian Navy Base, HMAS Creswell, in Jervis Bay (60 km to the south).

What to believe?

On top of the discrepancies between the *questionable* satellite data and reliable tide gauges, there is still the overriding problem about the sea level rise acceleration predicted by the computer models used by the IPCC. The models require acceleration to justify IPCC sea level rise scenarios that presently lie between 26 cm and 82 cm to the year 2100. In these models there is a false mathematical link between the rise in greenhouse gases since the middle of the 19th century and accelerating sea levels. *Unless sea level rise accelerates throughout the 21st century, the IPCC high sea-level scenarios are mathematically impossible, the computer models crash, and the climate position, espoused by the IPCC for the past 40 years, totally loses its credibility.*

Given the satellite problems, it is bizarre to have high confidence in the rising exponential sea level scenarios in IPCC documents and to describe them as the best fit (most likely) type curves for the next 100 years. It is an abuse of the English language to call them *precautionary*. Such terms imply that these projections have an empirical basis and reflect the way that sea level is actually behaving – characteristics the IPCC graphs are lacking. And, as if the IPCC projections were not bad enough, various scientific bodies and coastal authorities, in their desire to be vigilant and conservative, have added further increases in sea level to the IPCC scenarios as extra precautionary buffers.

The right choice

Any belief that the present satellite sea level data are superior to well-kept tide gauge data is totally misplaced. Regional planning should take into account that:

· There is evidence of steady sea level rise over the past 160 years, but contrary to the models used by the IPCC, there is no evidence for any long-term acceleration during that rise.

· There are regional tide gauges around the world providing excellent regional sea level rise data for the past 200 years, and each region should carefully examine its tide gauge records after making any corrections for local movements in the Earth's crust.

- Satellite derived sea level data come from a troubled satellite system that NASA wants to replace due to its data being at odds with long-term reliable tide gauges and having limited value. Readings of sea level rise rates from this satellite altimetry system are often two to three times above those of nearby reliable tide gauges.
- Reliable tide gauge data should be the basis for coastal infrastructure planning and coastal regional development.

Chapter 7.

Estimating coastal retreat

New lines in the sand

Many people in the world who live in low lying coastal regions are now being directly confronted with the climate debate when they are told by local authorities that their homes might be threatened by sea level rise. These authorities have to draft planning guidelines for coastal areas and to consider sea level rise. This process is a double-edged sword, as authorities risk legal action for allowing development in areas that are inundated by any sea level rise but, conversely, also risk litigation should their guidelines be shown to be based on erroneous assumptions.

For many in authority, this is the only time they will be directly affected by the climate debate. Most have little scientific background, but some will still venture a range of opinions, and local debates about sea level rise and climate change are often acrimonious.

In order to deal with this problem, normally central governments will set a possible range of future sea levels, and then instruct the local authorities to develop detailed regional guidelines that establish set-back lines from the ocean for future development and to forbid any type of construction in certain areas.

Local authorities normally employ environmental engineers to draft their regional coastal development policies. These people are normally well qualified to deal with erosional processes or with structural problems of sand and rock stability. However, when confronted with sea level change and the climate debate, they are often totally out of their depth. Explanations of climate in their reports are often based on *cut and paste* from the climate literature to make their reports look credible.

Tide gauges versus the satellite

An obvious problem facing those formulating coastal policies is the large divergence, between sea level rise data from satellite measurements and sea level rise data from well kept local tide gauges, as explained in the previous chapter. How can these engineers, who are not qualified scientists, argue with scientists who want them to adjust local tide gauge records upwards to agree with the satellite data? Most of these consultants do not know of criticism, even from NASA, about the doubtful value of the satellite sea level rise data. And, if they reject the satellite data in their reports, will they be accused of rejecting the IPCC's position in the climate debate and be labelled climate sceptics?

Engineers need to explain to regional authorities that the most accurate sea level data come from well kept local and regional tide gauges, not from a satellite system that even NASA wants to replace. A regional sea level policy can then be developed and that policy should be subject to review each decade.

The iniquitous Bruun Rule

When they are developing a regional policy, engineers are required to estimate what will happen to a coastline if the sea level rises by a certain amount. To do this, engineers often use a formula called the Bruun Rule.

This rule was developed in 1954 by Peter Bruun, a US Army Corps engineer, and it determines how a beach retreats or advances as sea level rises or falls. The Bruun Rule is based on the general profile of a beach. It has inputs such as sand grain size, beach slope angle and how far sand can be moved offshore. The calculations are based on forces working at right angles to a theoretically infinitely straight beach. The Bruun Rule assumes that, as sea level rises or falls, the beach will move higher or lower and that the dunes will move forwards or backwards, but the beach slope angle and the basic dune profile will remain the same. *In general terms, the formula estimates one metre of sea level rise causes 100 metres of coastal retreat.*

The Bruun Rule is easy to use and provides calculations that may impress local authorities, but recent field data from all over the world show that the Bruun Rule does not work, and only produces false results. Professor Andrew Cooper, who is the Chair of the Northern Ireland Marine Forum and Professor Orrin Pilkey, who set up the Program for the Study of Developed Shorelines (PSDS) in the US have jointly analysed the deficiencies of the Bruun Rule:

> *The Bruun Rule assumes that rising sea level always causes shoreline retreat... In nature many shore facies have been known to accrete (that is, grow seaward) even under rapid sea level rise... Several assumptions behind the Bruun Rule are known to be false and nowhere has the*

Bruun Rule been adequately proven; on the contrary several studies disprove it in the field. The Bruun Rule has no power for predicting shoreline behaviour under rising sea level and should be abandoned. (Cooper and Pilkey, 2004).

To further illustrate this argument, the Bruun Rule never considered the complex role beach vegetation can play as sea level changes. In many beaches there was increased vegetation in the foredunes that acted as an effective sand trap; a vegetation change that may have been a response to higher levels of carbon dioxide stimulating vegetation growth.

The author of this book lives in a small village of 2000 people, 200 kilometres south of Sydney, Australia. The local authorities commissioned engineers to draw lines on beach and cliff properties showing what would happen if sea level rose 90 cm. The engineers were paid handsomely and wrote lengthy, but largely incorrect reports. They were not geologists and did not know that whole areas of the coast in this region were tilted inland and that the rock platforms were tilted upwards one to two metres above sea level forming bulwarks against sea level rise. They wrongly presumed the rock platforms were at sea level, and so the cliffs behind them were immediately susceptible to any sea level rise. The engineers did not get any sound geological advice.

Around the beaches these engineers made further mistakes. Local residents, whose houses had been built on the actual sand dune system, told regional authorities that there was now more sand in front of their houses than 40–60 years ago. The council environmental engineers were quick to tell them that they were not scientists. To resolve the problem this author superimposed aerial photos taken after World War II over other aerial photos taken at various times up to 2008. The locals were correct. Their beaches were stable and in some places the beaches were advancing seaward. The reason was obvious looking at comparative photos using a CAD computer program. There had been an explosion of foredune vegetation during the last 60 years; maybe due to extra carbon dioxide stimulating that plant growth. So over the years the increased vegetation had trapped millions of tonnes of sand, and when this happened the coast in this region was stabilised as sea level rose and actually moved seaward in many places. This movement was quite opposite to simplistic linear predictions based on the Bruun Rule. An example of such movement can be seen at the beach near Kinghorn Point, southern New South Wales, Australia (Diagram 7.1).

Diagram 7.1: Complex shoreline behaviour in the foredune at Kinghorn Point, New South Wales, Australia (Photo, H. Brady, 2012).

The vegetation in the foredune is flourishing and trapping so much sand that this beach, like others in this region, is actually advancing seawards as sea level rises. When there are storm surges the foredunes on these beaches are destroyed, but then the foredunes re-establish themselves and grow seaward again.

Unfortunately, the Bruun Rule has been used by environmental consultants for the past 60 years, and has *consequently decreased beachfront property values all over the world by trillions of dollars.*

The complexity of shoreline movement in *low lying Pacific Islands*, such as Kiribati, Tuvalu and others, shows the deficiency of the Bruun Rule. As the sea level has risen, not all of these coastlines have retreated. Rather, some coastlines have been *stable*, some have *advanced* seaward, and *only a few have experienced coastal retreat.* This variability is contrary to the Bruun Rule.

In 2010, Dr Arthur Webb, a Fellow of the Australian National Centre for Ocean Resources and Security (ANCORS), and Professor Paul Kench of the University of Auckland, New Zealand, carefully analysed aerial photos of shorelines in this region taken over a 61-year period. They comment:

Firstly, islands are geomorphologically persistent features on atoll reef platforms and can increase in island area despite sea-level change. Secondly, islands are dynamic landforms that undergo a range of physical adjustments in responses to changing boundary conditions, of which sea level is just one factor. Thirdly, erosion of island shorelines must be reconsidered in the context of physical adjustments of the entire island shoreline as erosion may be balanced by progradation on other sectors of shorelines.

... for 86% of islands, 43% remained stable or 43% increased in area (over the timeframe of analysis). Largest decadal rates of increase in island area range between 0.1 to 5.6 ha. Only 14% of the study islands exhibited a net reduction in island area (Webb and Kench, 2010).

The above research by Webb and Kench is ignored by local politicians, even though it is based on solid data. Unfortunately, peoples in these islands are being classified as the first *climate refugees*. The same narrative is being used by politicians and spokespeople for other low lying islands, such as the Maldives in the Indian Ocean. For all such low lying islands, the blame for rising sea level is placed squarely on industrialised nations for increasing greenhouse gases over the past 150 years. As a consequence large financial compensation is being demanded in speeches at the United Nations and at group meetings such as the Pacific Islands Forum. Climate alarmists also frequently turn their attention to low lying Bangladesh – an area very vulnerable to storm surges. Yet the main change in this region is a shallowing of the Bay of Bengal as the Ganges adds more sediment to its seafloor every year (Carter 2010, p97).

Storm surges

Apart from sea level rise caused by climate change, high sea levels occur during storms associated with the low pressure systems of cyclones and hurricanes. Such systems cause a local rise in sea level that can last for a few days. The lower the pressure, the higher the sea level rise, and that rise can be well above the IPCC sea level rise scenarios for the next 100 years. These rises in sea level are more severe as one approaches the tropics and are less severe in the mid-latitudes. For example, in October 2012, a super-storm, Cyclone Sandy, crossed the eastern coast of the United States. Due to extreme low pressures, the effective sea level rise at New York was above four metres, and parts of New York and its subways and the nearby coastline were flooded (see Photo in Diagram 7.2).

Diagram 7.2: Photo of the storm surge inundating the Atlantic City region just south of New York, October 2012.

Sea level naturally rises in areas of low atmospheric pressure. The very low atmospheric pressure of Cyclone Sandy allowed sea level to rise around 4 metres and the ocean flooded into New York City. Billions of dollars are being spent on future climate research, and often little money is spent to lessen the effect of storm surges in vulnerable areas.

Therefore, it is absurd if regional authorities spend millions of dollars formulating policies to deal with future IPCC sea level rise scenarios, but do little to develop strategies to deal with storm surges that occur regularly and do considerable damage. Even though the highest sea level rise will correspond with the low pressure eye of the storm, such storms generate long amplitude waves and cause coastal erosion over much larger areas.

Regional coastal policy

While precautionary coastal development policies to deal with sea level rise are important, there is no simple linear relationship between sea level rise and shoreline history. Nonlinear coastal behaviour means that it impossible to have a uniform coastal policy in any large country. Such policies should be based on:

- *The actual coastal changes over the past century:* data from surface surveys and data from the aerial photo history of a region should be collected. A careful examination of surveyor records and photographic history should show how a local coastline is behaving. In some regions the coast will be stable, as sea level rises, in other regions the coast may be retreating landwards or even advancing seawards.

- *Regional tide gauge sea level rise data:* if NASA satellite sea level data are at odds with data from regional tide gauges, the regional data should be used and not adjusted upwards to data gathered by the present satellite system that NASA wants to replace. NASA acknowledges the clash between excellent long-term tide gauges and data from the Jason satellite.

- *Re-examination of previous coastal policies:* local authorities need to re-examine and to revise any estimates of coastal retreat due to sea level rise, especially if those estimates were calculated using the defective Bruun Rule.

- *Data from previous storm surges:* an analysis of previous local storm surges should be more than adequate to establish enough data for regional authorities to formulate strategies to best mitigate storm surge effects.

Because of regional differences, a uniform national sea level policy is not possible. Each region needs to intimately understand its own coastline. Local observational data more accurately reflects the nonlinearity of the coastal processes and the singularities and idiosyncrasies of each region.

Chapter 8.

The poles will drown us

Heroic adventure and mythology

The polar regions have always attracted media attention, presenting the allure of unknown, far away, lands. The newspapers in London, Berlin, Oslo and Paris funded the first expedition led by Amundsen to reach the South Pole in late 1911. Such was the public interest in the expedition, Amundsen sailed north to Tasmania where he knew he could telex the news of his success to the whole world. Rather than releasing news of his triumph immediately, he crept in disguise into Hobart, booked into the Hadleys Hotel, and then telexed his brother in code to inform his newspaper backers. Similarly, one of the main backers of the ill-fated expedition which left from Cardiff for the South Pole, led by British Navy Captain Randolph Falcon Scott, was The Western Mail – the leading newspaper in Wales.

The media's interest in the polar regions has not abated, but now the stories are about pack ice disappearing, and ice sheets falling into the sea causing the sea level to rise at an alarming rate. These stories originate from scientists who believe these processes can happen quickly due to global warming and such yarns are then dramatised by the media. However, geological history tells us that even in the much warmer mid-latitudes, ice sheets that advanced southwards as far as 45°N, take at least a few thousand years to melt. Indeed, *we have no evidence of any rapid demise of polar ice sheets due to their almost permanent sub-zero surface temperatures.*

The ancient writers of mythology took poetic licence and condensed history by shortening events that took thousands or millions of years to a few days. So is present day climate science, once fuelled by the media, creating a new mythology with talk of the West Antarctic Ice Sheet or the Greenland Ice Sheet sliding quickly into the sea?

The key questions in polar regions are:

· What are the dynamics controlling areas covered by pack ice?
· What are the dynamics controlling the volume of ice sheets?
· Is climate change in polar regions as dramatic as inferred in the media?
· Is climate change in polar regions linked to recent global warming?

The Arctic pack

Our knowledge of the Arctic pack ice is limited to the past 200 years of exploration. This ice covers the ocean at the North Pole where there is no land mass. This pack ice forms as the surface waters of the ocean freeze. Since that ice comes from the ocean, it does not affect world sea level. Danish maps of the Arctic pack ice exist from the 1920s and satellite monitoring of the Arctic pack ice began in 1979. The Arctic pack ice varies in area from around 16 million sq km in winter to 6 million sq km in summer. Since some of this pack ice melts in summer months, the winter pack ice is a mixture of new ice formed each winter, and older ice from previous winters.

The Danish maps document:

· A warm Arctic period in the 1920s and 1930s;
· A cooling trend from 1938 onwards with the Arctic pack ice extent increasing in extent until the 1980s.

Modern satellite records show a dramatic decrease in the summer extent of this pack ice from 2000 until it reached a record low of 3.41 million sq km in 2012. But, since 2013, this trend may have reversed as the Arctic summer ice cover rebounded in 2014 to 5.02 million sq km. The data may indicate the beginning of a colder cycle in the Arctic, and an increase in pack ice cover in the coming decade.

The Antarctic pack

The Antarctic pack ice is separated from the world's oceans because the largest ocean current in the world (the Antarctic Circumpolar Current) flows clockwise around Antarctica and effectively isolates the Antarctic continent and the pack ice in the Southern Ocean. Unlike the Arctic, a greater percentage of the Antarctic pack ice melts in summer, so this southern pack ice is mainly new ice. As summer ends around late February this ice can form at over 100,000 sq km a day until it eventually covers over 20 million sq km of ocean by late October. It then decreases to around 3 million sq km by late January.

The first explorers to confront this pack ice were the American whalers from Boston and San Francisco who ventured south in the first few decades of the 19th century. The British had placed a tax on whale oil sold in London and this undermined the whale oil market for American whalers. It was now more attractive to hunt seals and sell seal skins to Asia. Their seal hunts concentrated on the sub-Antarctic islands, and it was only on rare occasions they made any headway through the Antarctic pack ice.

A Russian admiral, Fabian Gottlieb Thaddeus Bellinghausen, made his way through the pack ice in 1820 and was the first to sail along the West Antarctic coast, but this mountainous region was not a gateway to the interior. It was not until January 1841 that Sir James Ross, with specially strengthened wooden ships, weaved his way through the summer pack to enter what is now called the Ross Sea and sailed along the mainland coast. This famous voyage established a route that later explorers, such as Borchgrevink, Amundsen, Shackleton and Scott, used to reach the mainland and for some to attempt to reach the South Pole.

Because of the size and remoteness of Antarctica, mapping of the Antarctic pack ice was patchy until satellite monitoring began in the 1970s. Since then the Antarctic pack ice has increased in a linear trend for 40 years; *a trend totally opposite to that in the Arctic pack ice* (Diagram 8.1).

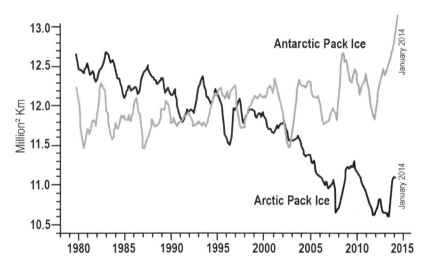

Diagram 8.1: Arctic and Antarctic pack ice extent 1978–2014 (National Sea Ice & Data Centre, US).

The maximum surface extent of the Arctic and Antarctic pack are out of sync. The maximum extent of the Antarctic pack ice has been steadily increasing while maximum extent of the Arctic pack ice has been decreasing, although there was a small rebound in 2014 and 2015. In the Antarctic, the pack ice sits in a region south of the huge Circumpolar Current and is not affected by currents from the Pacific, Indian and Atlantic Oceans. Its aerial extent has only been mapped in recent years using satellites. The actual mechanism controlling this pack ice is not known and the reasons behind the steady increase in the maximum extent of Antarctic pack ice over the past 37 years are not understood.

Ocean currents and pack ice

The Arctic pack ice can be influenced by ocean currents from either the Pacific Ocean through Bering Strait between Alaska and Siberia, or from the North Atlantic Ocean by the currents flowing along the east or west coasts of Greenland. There is strong evidence that the ocean currents in the North Pacific and Atlantic Oceans alter the amount of Arctic pack ice. There is also evidence that changes in wind patterns can affect the amount of old ice that is blown out of the Arctic into the Atlantic Ocean (Diagram 8.2).

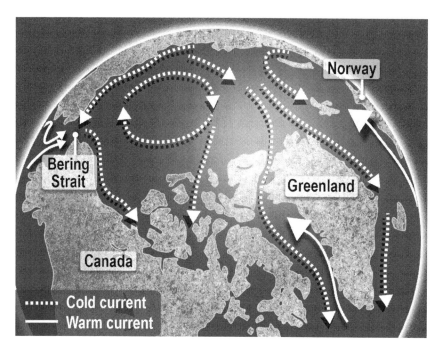

Diagram 8.2: This diagram from Wikipedia shows ocean currents around the Arctic.

Warm and cold currents, in periodic cycles, may be responsible for variations in the Arctic pack ice. Warm currents can come through Bering Strait near Alaska (in one phase of the Pacific Decadal Oscillation), and also along the west coast of Greenland and along the eastern coast of Norway in the Atlantic Ocean. The extent of the Arctic pack ice declined strongly in the 1920s and 30s, then increased, and then declined between 1990 and 2012. It may be slightly increasing at the present time. Some villages in Greenland lie on the west coast because a warm current hugs that coast.

We have already mentioned the switch in temperatures in the North Pacific Ocean called the Pacific Decadal Oscillation (PDO). A 30-year phase of this 50–60 year cycle can explain the decrease of Arctic ice in the 1920s and 30s, as warm waters from the North Pacific Ocean flowed through Bering Strait into the Chukchi Sea in the Western Arctic. In the other 30-year phase, no warm water will enter Bering Strait.

There is a similar process in the North Atlantic Ocean. On different time scales, Atlantic currents along the south-west of Greenland and in the eastern North Atlantic, can vary the amount of warm water entering the Arctic Ocean and can also affect the extent of the Arctic pack ice in the Barents and Kara Seas that lie along the northern Russian and Siberian coasts.

Future changes in the Arctic pack ice are quite possible. The driving force is not present atmospheric temperatures but the heat in ocean currents accumulated over many years. If the present global warming trend continues, the amount of Arctic pack ice may continue to decrease each time the PDO and northern Atlantic ocean currents push warm water into the Arctic.

In this nonlinear system, there is one critical, ticking time-bomb in the Arctic. Heavier, salty, warm water from the North Atlantic lies under lighter, less salty but colder surface waters. If overturned this warm water could melt all the ice in the Arctic, but this anomalous event may never occur. On November 20, 2014, Professor Andreas Muenchow, from the University of Delaware, posted this comment on his internet site *Icy Seas*:

> ... *As this warm water moves counter-clockwise around the Arctic Ocean to the north of Siberia and Alaska, it subducts, that is, it is covered by cold water that floats above the warm Atlantic water. So salty but heavier, warmer water is lurking under the fresher, lighter, colder water near the surface of the Arctic. If overturned this warm water could melt the Arctic ice within days!*

Statements about the dramatic decrease in Arctic pack ice have been made by popular climate catastrophists such as Al Gore, an ex vice-president of the United States. He promoted ideas developed in 2007 by Professor Wieslav Maslowski of the Naval Postgraduate College of California who had argued the Arctic could be ice-free by 2013 and that *"you can argue that our projection of 2013 could be too conservative"* (BBC News December 12, 2007). Al Gore presented this theme when he received the Nobel Peace Prize for his climate work:

Last September 21, 2007, as the Northern Hemisphere tilted away from the Sun, scientists reported with unprecedented distress that the North Polar ice cap is 'falling off a cliff'. One study estimated that it could be completely gone during summer in less than 22 years. Another new study, to be presented by U.S. Navy researchers later this week, warns it could happen in as little as 7 years.

This forecast became a major embarrassment for Al Gore. It looked on track in 2012 as the summer Arctic pack ice decreased to just over 3 million sq km from its old average around 6 million sq km but by 2014 it had bounced back to over 5 million sq km.

All these variations in the Arctic pack ice are evidence of climate change; however, they do not directly affect world sea levels. The overriding conundrum has been the behaviour of the Antarctic pack ice. Contrary to the predictions of climate models, this pack ice has increased in areal extent for 40 years, and not decreased like that in the Arctic (Diagram 8.1). Such behaviour underscores the nonlinear nature of climate, and the inherent limited value of climate models. Being cut off from the world's oceans by the Circumpolar Current, the Southern Ocean has not seen a significant intrusion of warmer waters from the Indian, Pacific and Atlantic Oceans except perhaps around the West Antarctic Peninsula jutting out towards South America.

The polar land masses and ice sheets

Large ice sheets, two to four km thick, cover the island of Greenland in the North Atlantic Ocean, and the Antarctic continent in the South Pole region. Their weight depresses the land, in some places below sea level. If melted, they would raise world sea levels around 70 metres. Talk of melting these ice sheets interests the world's media. The land of Greenland covers 2.1 million sq km and scientists calculate that the melting of the Greenland Ice Sheet could raise world sea level around six metres. However, the Antarctic land mass is seven times larger than Greenland and covers 14 million sq km. Here the large East Antarctic Ice Sheet has enough ice to raise world sea levels 55 metres, and the West Antarctic Ice Sheet, on the peninsula jutting out towards South America, has enough ice to raise world sea levels by about six metres.

Inside these continents there are large areas of moving ice, much like slow rivers. These features are called ice streams because the moving ice is bounded by ice, not by rock or mountains. These ice streams can be 100 kilometres wide, two kilometres deep and hundreds of kilometres long.

There are also glaciers moving ice in coastal valleys between coastal mountain ranges in Greenland and Antarctica. Often, when these glaciers reach the ocean, they can form large floating tongues of ice that can extend kilometres into the sea. The ends of these tongues break off and form large icebergs. But in some areas these glaciers and ice streams can coalesce to form huge areas of ice shelves that can sit on or float over the ocean floor. The largest ice shelf is the Ross Ice Shelf in Antarctica that covers 487,000 sq km; the size of Spain!

Greenland and ice loss

In Greenland, there is evidence that ocean currents are causing the retreat of coastal glaciers by undercutting the glacial tongues that float in fjords along the Greenland coast. This undercutting is like a hidden cancer; sometimes not initially obvious from surface photos. Professor Andreas Muenchow, notes on his internet site *Icy Seas*:

> While the exact contributions and details of ocean melting vary from glacier to glacier, little doubt exists that the ocean's heat and currents contribute to retreating glaciers. And yet, nobody really knows how the heat from the deep Atlantic Ocean 1. crosses shallow and broad continental shelves to 2. enter the coastal fjords, and to 3. reach the glaciers.

Apart from ocean currents undercutting Greenland glaciers, the surface of the Greenland Ice Sheet shows evidence of dynamic change. During the summer there are melt days where the surface is an icy mush and in many places small lakes form and these sometimes drain quickly into rifts within the ice sheet. If such rifts, over two kilometres deep, are filled with this lake water, the water pressure deep in the rift can create further cracks in its sides; this process is much like the 'fraccing' technique used in the oil industry to break open rocks at depth. If such rifts are near the coastal glaciers, this under-ice water can flow under glaciers and speed their path to the sea. Professor Petr Chylek, of Dalhousie University in Halifax, Nova Scotia, has looked at the present relationship between coastal temperatures and mush days on the ice sheet and argued that these processes must have been also very active in the 1930s (Diagram 8.4).

The undercutting of glacial tongues by the ocean, plus the processes by which surface waters can penetrate the Greenland Ice Sheet near the coast has meant that many Greenland glaciers flow at rates exceeding five kilometres a year. The dynamic Jacobshavn Glacier has been flowing around 18 km/year; the fastest in the world. Other ice losses are due to high winds evaporating ice (ablating) even at low temperatures. These losses have to be balanced against ice gains due to increased snowfall near the coast when low pressure atmospheric systems bring moisture over the plateau where it is dumped as snow. Dramatic evidence of this snowfall comes from the crash of six P-38 fighters and two B-17 bombers on the Greenland Ice Sheet in 1942. *One of the planes, now called Glacier Girl, was recovered from the ice sheet in 1992 and it was buried under 81 metres (268 feet) of snow; a height gain in this region of 1.6 metres per year over 50 years!*

The Earth's oceans are so large, it takes about 360,000 gigatonnes of ice to raise world sea level one metre; and one gigatonne (Gt) is 1000 million tonnes. The satellite data indicate that the annual ice loss from Greenland varies from year to year. Recent estimates have been around 267 Gt in 2007, 570 Gt tonnes in 2012, 280 Gt in 2013 and six Gt in 2014. While these changes in Greenland Ice Sheet look dramatic, they are only causing recent sea level rise up to 0.7mm per year (7 cm/100 year). And Greenland's average annual contribution to sea level over the last 100 years would be lower.

It is evident that the Greenland Ice Sheet is the most vulnerable to climate change and that, at the present time, its contribution to sea level rise exceeds any contribution from the Antarctic Ice Sheets. However, there is no evidence of any immediate catastrophic collapse. The spectacular meltwaters documented today must have occurred during the Medieval Warm Period when the Vikings lived in Greenland a thousand years ago; there was no collapse then. *During that time the Vikings had cattle and they could easily bury their dead because there was no permafrost in southern Greenland, as there is today.* If global warming continues, the Greenland Ice Sheet will be the most vulnerable to change, but not overnight. There is evidence it has collapsed in the past million years; fossil tree pollen, retrieved from drill cores to the base of the ice, show that after it collapses Greenland is re-forested.

Antarctic ice losses and gains

Antarctica is cut off from the world's oceans by the Antarctic Circumpolar Current that flows clockwise around the whole continent. Besides isolating the Antarctic mainland, this current connects with warmer currents from the Indian, Atlantic and Pacific Oceans (Diagram 8.3). There can be warmer swirls in this Antarctic Circumpolar Current from these warmer currents, and such warmer water could reach the Antarctic mainland along the West Antarctic Peninsula. Some present undercutting of glaciers on this peninsula may be caused by this effect (a similar process to the undercutting of glaciers along the Greenland coast in the Arctic, as already described).

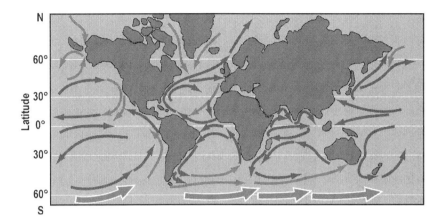

Diagram 8.3: The currents in the Earth's oceans (Wikipedia).

The diagram shows how ocean currents along the eastern coasts of South America, South Africa and Australia flow southwards and connect to the Circumpolar Current that flows clockwise around Antarctica. This Circumpolar Current is the largest current on Earth and it flows at 100–150 Sverdrups/second (one Sverdrup = 1 million cubic metres). This current effectively isolates Antarctica from the rest of the Earth.

The Antarctic mainland is a roughly circular continent with a mountainous peninsula jutting out towards South America (Diagram 8.4). The land between the Ross and Weddell Seas is so weighed down by the West Antarctic Ice Sheet that is up to 4 km in thickness and often depressed below sea level.

There are many research stations of various sizes on the Palmer Peninsula or on the nearby sub-Antarctic islands. The main ones belong to Britain, the US, Chile and Argentina. In the political rush to have some claims to the riches of Antarctica, many other small stations are being built by other nations.

The bulk of the continent is covered by the large East Antarctic Ice Sheet. It has so buckled the Earth's crust that the Ice Sheet effectively sits in a saucer with its coastal fringe rimmed by mountains. The most spectacular mountain chain is called the Transantarctic Mountains. Along the coast are many scientific research stations. Important stations are McMurdo (US), Dumont d'Urville (France), Mirny (Russia), Casey and Davis (Australia), Showa (Japan). In the continental interior lie the important US base at the South Pole and the Russian station named Vostok – the coldest place on earth.

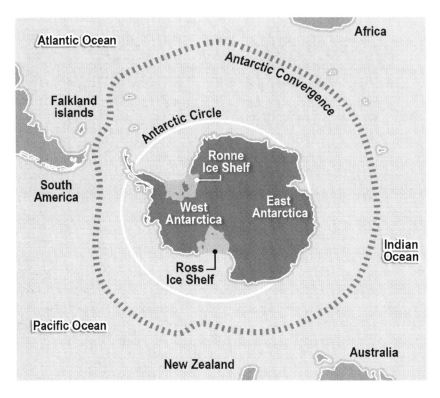

Diagram 8.4: The Antarctic continent of 14 million sq km, nearly twice the size of Australia.

The meeting place between the Pacific, Indian and Atlantic Oceans with the Southern Ocean is the Antarctic Convergence. This is the largest ecological barrier on Earth. Over a distance of only 50 km the water temperature drops from 5.5°C to 2.8°C. This completes the isolation of Antarctica. The Ronne Ice Shelf and the Ross Ice Shelf, each nearly the size of Spain, are fed from the West Antarctic Ice Sheet and from the East Antarctic Ice Sheet.

There has been a warming trend along the West Antarctic peninsula and in sub-Antarctic Islands, and some coastal stations, while the deep heart of the Antarctic continent has cooled. The mean temperature at the Signy-Orcadas Island in the South Orkney group, from station records, has risen by 3°C over 100 years, but the actual South Pole temperatures have cooled slightly over the past 36 years. The coastal stations have mean temperatures between −9°C and −15°C, and the high altitude stations in the interior, have mean temperatures around −50°C.

The best documented ice loss on the Antarctic mainland has been from some coastal glaciers along this West Antarctic peninsula. For example, the Pope, Smith and Thwaites glaciers retreated 31 km, 35 km and 14 km respectively between 1992 and 2011. Eric Rignot, from the University of California, and Ian Joughin from the University of Washington, were the principal authors of two papers on the retreat of the Pine Island and Thwaites glaciers. These authors propose a retreat of the West Antarctic Ice Sheet in the next 100–400 years, rather than a quick collapse (Rignot et al., 2014, Joughin et al., 2014).

A panicky media placed these reports in context with an old theory, proposed in the 1970s, that the West Antarctic Ice Sheet is in danger of imminent collapse and is *"the gorilla in the sea level closet"*. On hearing the news on the retreat of these West Antarctic glaciers, Jerry Brown, the Governor of California suggested moving the Los Angeles and San Francisco airports.

Apart from the coastal glaciers just mentioned, studies are now focusing on the continental interior covered by the West Antarctic and the East Antarctic Ice Sheets. These ice sheets are up to four km thick and their interior plateaus, well above sea level, have mean surface average temperatures around −50°C.

Ice streams move ice from these plateaus toward the ocean. Because their surfaces are so cold, the draining of these ice sheets by ice streams is not controlled by surface air temperatures or by recent global warming. The depth of the ice and plastic flow within the ice are responsible for their movement. There are no 'mush' days, as on the Greenland Ice Sheet, and no surface lakes that could find their way through rifts to its base. There are however, many deep sub-surface lakes within both ice sheets. The dynamic processes forming these sub-ice lakes are far from understood. Sometimes one sub-surface lake can drain and join another at a different pressure and the height of the ice sheet surface in that area will adjust. If one of these larger lakes is near the coast then it could lubricate glacial flow in that area.

Ice streams on the West Antarctic ice sheet flow towards the Ross Sea or in the opposite direction to the Weddell Sea. A 2009 paper from the Rignot group shows that the main ice streams, the Mercer and Whillans

Ice Streams, have decelerated slightly between 1997 and 2009. These ice streams account for 80% or more of ice movement from the West Antarctic Ice Sheet. Where these large ice streams pour into the sea, they have created the large Ross Ice Shelf in the Ross Sea region and Filchner-Ronne Ice Shelves in the Weddell Sea region. These are huge areas of ice, in places floating on the sea, and in others grounded onto the sea floor. The Ross Shelf covers 487,000 sq km and the Filchner-Ronne Ice Shelves 430,000 sq km. In places they are well in excess of 500 metres thick. The author of this book was a field geologist for the Ross Ice Shelf drilling project when this Ice Shelf was drilled for the first time in 1977 and 1978. At the drilling location the Ice Shelf was 420 metres thick. Live krill and small fish were noted in the 237 metres water column under the Ice Shelf even though the drill site was 500 km from the open ocean.

Recent data from the CryoSat satellite showed a stable East Antarctic Ice Sheet and some ice loss from the West Antarctic Ice Sheet. The media exaggerated the significance of this West Antarctic data and reported that the data *"surpassed scientists' worst fears"* and was enough *"to raise world sea levels by 0.45 mm every year"*. However, a little excursion into basic mathematics would not hurt! Taking this CryoSat data as factual and as the best available (which it isn't), and extrapolating the data, would indicate that West Antarctica would only contribute about 4.5+ cm of sea level rise during the next 100 years; so much for ill-informed media panic!

The analysis of ice losses and gains on polar ice sheets uses data from a number of satellites and is quite complicated. Atmospheric circulation can bring moisture from low pressure systems over the ice sheets where it is precipitated as snow. The British Antarctic Survey has inspected ice cores and reported a dramatic increase of snowfall in West Antarctica in the 20th century (Thomas et al., 2015). Such snowfall increase has to be offset against ice loss into the ocean from coastal glaciers and ice streams, or from ice loss due to the evaporation of ice by high winds (ablation). No water is running into the sea as the surface temperatures of the Antarctic ice sheets are so low that no melt water is formed.

In November 2015, Dr Jay Zwally, Chief Cryospheric Scientist at NASA's Goddard Space Flight Centre and Project Scientist for the Ice Cloud and Land Elevation Satellite, reported in *The Journal of Glaciology*, that the large Antarctic East Ice Sheet is growing and more than compensating for any ice loss from the West Antarctic Ice Sheet. The increase is due to increased snowfall. Dr Zwally reported that analysis of satellite data from 2003 to 2008 showed a net gain of 82 billion tonnes of ice per year during that period; a clear suggestion that the Antarctic ice sheets could contribute little to sea level rise this 21st century (Zwally et al., 2015). This latest report contradicts a 2013 IPCC report that indicated a net ice loss in the Antarctic of 147 billion tonnes per year between 2002 and 2011. However, Dr Zwally points out that his study is a report of what is actually happening now, and is not a prediction for the future.

Spanner in the works.

Interesting evidence from West Antarctica contradicts comments of those promoting recent global warming as *'unprecedented'*. Scientists from the British Antarctic Survey analysed an ice core taken by them on the Bryan Coast, Ellsworth Land, adjacent to the Bellingshausen Sea in 2010/11. This was one of a series of ice cores, and it provided a 308-year temperature record. There were various volcanic ash layers in the ice core that could be dated with an error margin within one year. These ash layers were from various eruptions of Chilean volcanoes in the 18th, 19th and 20th centuries. The results are self-explanatory:

> *The record shows that the region has warmed since the 1950s, at a similar magnitude to that observed in the Antarctic Peninsula and central West Antarctica; however this warming trend is not unique. More dramatic isotropic warming (and cooling trends) occurred in the mid-nineteenth and eighteenth centuries, suggesting that at present, the effect of the anthropogenic (human) climate drivers at this location has not exceed the natural range of the climate variability in the context of the past ~300 years...*

> *However, the recent isotropic warming trend is not the largest in the 308 year record. Larger 50-year warming trends occurred in the middle to late eighteenth century (+4.1% per decade, 1740–1789) and the mid-nineteenth century (+3.8% per decade, 1839–1888) with several equally large cooling trends. Overall there is no significant trend in the average deuterium record since 1702 A.D.* (Thomas et al., 2013).

So how can scientists say that the recent global warming in the polar areas was unprecedented when there was a stronger warming trend 250 years ago in West Antarctica? And this warming occurred before any rise in carbon dioxide levels!

Distinct layers in this core show 50-year temperature cycles. This is remarkably similar to the 60-year pause-warm cyclicity over the past 150 years (see Chapter 5 of this book). Many scientists have treated the pause-warm patterns in the past 150 years as unusual, simply due to the fact that nothing was known about these small variations in the past. This new data means any such view is wrong, as the new data links mid-18th century climate change to the present. And contrary to this historical evidence, computer climate models do not show a step-like pattern in temperature movements over the past 300 years, and are not even close to mimicking climate change in the real world.

No surprises

It is not surprising that high latitude and high altitude glaciers have retreated. Such glaciers have always been very sensitive to climate change. Today, pictures of dramatic climate change in high latitude areas or in high altitude mountains are being used to prove we are on the verge of catastrophic climate change. Indeed, a visit by President Obama to a retreating glacier in Alaska in September 2015 was used to present this very misleading picture of global climate change. In truth, rapid dramatic climate change in high altitude mountains or in polar areas, unlike mid-latitude and tropical areas, is simply proof that those areas have always been extremely sensitive to even the smallest changes in climate – a fact pointed out over 100 years ago by Milutin Milankovitch and James Croll when they examined the reasons for the cyclical nature of ice advances during an Ice Age.

It is interesting to note that *in Greenland the strongest warming trend in the 20th century was in the 1930s as shown by the temperature records of the settlements at Godthaab Nuuk and Egedesminde on the west coast* (Diagram 8.5 a & b). These graphs are interesting. Melt days occur on the surface of the Greenland Ice Sheet on fine days when very still conditions lead to a situation where large areas become a *mushy melt.* These melt days are now recorded. It should not be thought that this water flows to the sea, but it can form lakes and disappear down rifts to the base of the ice sheet. At the present time, the main ice losses are from glacial flow and the undercutting of coastal glaciers by warm ocean water along the coastal margin.

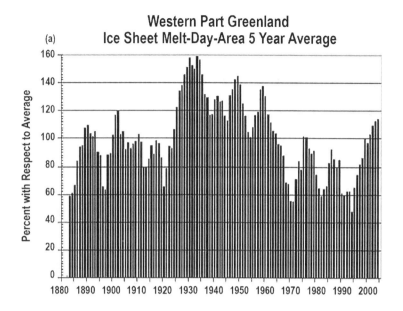

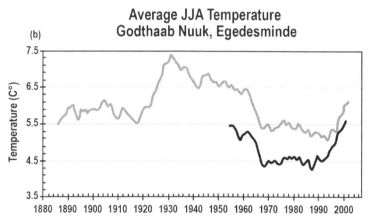

Diagram 8.5: a) and b): The temperature record of two Greenland villages is used to reconstruct mush days on the Greenland Ice Sheet (Chylek et al., 2006).

The top diagram is an estimate of surface ice melt days on the Greenland Ice Sheet based on the lower diagram of temperature records for two west coast Greenland villages, Godthaab Nuuk and Egedesminde, between 1880 and 2005. The bottom graph shows that local temperatures on the West Greenland coast in the 1930s were higher than those of today. This West Greenland data is consistent with the 1912–2010 temperature record of the island of Svalbaard in the Greenland Sea where the warming between

1920 and 1925 was the fastest recorded anywhere in the 20th century. An analysis of the temperatures records for 37 Arctic stations and seven sub-Arctic stations also showed the highest temperatures up to 2005 were in the 1930s (Przbylak, 2000).

A last look

Looking at the pack ice in both hemispheres, there is no obvious correlation between rising carbon dioxide levels, the cyclical changes in the Arctic pack ice and the gradual, almost linear, increase of the Antarctic pack ice to record levels over the past 40 years. This lack of synchronicity is typical of a nonlinear system. Climate models wrongly forecasted the early demise of Arctic pack ice, and no climate model foresaw the steady increase in Antarctic pack ice over the past 40 years as carbon dioxide levels rose. Nor can any climate model explain warming in the mid-18th century equal to or greater than the recent warming trend between 1975 and 2000.

The latest NASA figures for Antarctica indicate it has made no net contribution to world sea level rise this century as the growth of the East Antarctic Ice Sheet has more than compensated the ice losses from the West Antarctic Ice Sheet. The balance to ice losses has been increased snowfall over the much larger East Antarctic Ice Sheet (Zwally, 2015). The ice loss figures from Greenland fluctuate wildly, but presently indicate a contribution to world sea level rise of around about 7 cm/100 years. Even allowing for some expansion of the ocean and some increase in Greenland ice loss, the IPCC high sea level scenarios of 28–82 cm to the year 2100 remain mathematically impossible as they depend on a significant acceleration of net ice losses from Greenland and Antarctica – an acceleration not shown in the data.

Monitoring of the Antarctic and Greenland ice sheets is important because, at the end of the day, these ice sheets are the only areas that could supply large volumes of fresh water to raise world sea levels significantly.

Chapter 9.

The greenhouse effect

The jig-saw puzzle

The spectre of rising carbon dioxide levels is driving all the panic about accelerating temperatures and accelerating sea level rise, and accelerating frequency of severe storms. However, global temperature and global sea levels began to rise 300 years ago during the 18th century, well before carbon dioxide levels started to increase 150 years ago. Despite the panic, global sea levels are not accelerating and storm frequencies show no rising patterns. At the present time, greenhouse gas levels will continue to rise due to further industrial development and the ongoing significant increases in the human population.

The role of greenhouse gases cannot be considered in isolation from other factors that influence the Earth's climate. As mentioned in Chapter 2, there have been ice ages on Earth when greenhouse gases were over ten times their present level (Diagram 2.1). An in-depth examination of the role greenhouse gases have played and will play in the climate system shows that there are limitations to the greenhouse effect and that the role of each greenhouse gas is just one piece in a very complex jig-saw.

The main game

Without an atmosphere, the Earth would be a barren, cold, inhospitable planet. The Earth's atmosphere is mainly warmed by conduction of heat as air contacts the Earth's surface that has been heated by shortwave radiation from the Sun. Apart from conducting heat from the Earth's surface, the atmosphere also absorbs some energy reflected from the Earth as longwave radiation. The temperature of different surfaces, such as water, rock, vegetation, soil, snow and ice, determines the wavelength of the longwave radiation reflected into the atmosphere. Surprisingly, only certain atmospheric gases selectively absorb some of this energy. Such absorbed energy is re-emitted to further warm the Earth's atmosphere and this warming is known as the *greenhouse effect*. Most of the greenhouse effect takes place in the lowest two kilometres of the atmosphere; a region known as the lower troposphere.

The warming of the atmosphere by the Earth's surface and the greenhouse effect would not be effective without strong convection currents of air that move heat upwards and sideways. These convection currents are especially strong in the hot tropical regions as they help distribute heat towards the poles. This movement of heat polewards is an important factor in the Earth's energy budget, and it is helped by the spinning of the Earth. The Earth is like any rotating body that throws objects sideways. In this case the faster spinning atmosphere at the equator is deflected and thrown polewards.

Sponges in the sky

Leaving aside water vapour content (which varies tremendously by location and temperature), 99.96% of the gases in the atmosphere are nitrogen (78.08%), oxygen (20.95%) and argon (0.93%); none of which contributes to the greenhouse effect.

The greenhouse gases are water vapour, carbon dioxide, methane, nitrous oxide, and ozone.

The origin and distribution of the greenhouse gases are as follows:

- Water vapour comes from the evaporation of the surface waters of the oceans, rivers and lakes, from the evaporation and melting of snow and ice, and from the evaporation of ice and snow when high winds turn snow or ice straight into water vapour (ablation). In the atmosphere, water vapour decreases as the air gets colder and is virtually absent at an altitude of 10 km.

- Carbon dioxide comes from many sources such as volcanoes, the erosion of limestones, the combustion of fossil fuels, the burning of plant material and the outgassing of carbon dioxide from the ocean. Animals breathe in and combust oxygen, and then breathe out carbon dioxide. On the other side of this cycle, plants take in carbon dioxide from the air to make their basic food and then breathe out oxygen. Carbon dioxide does not easily freeze and so it stays well mixed to an altitude of 80 km in the upper atmosphere.

- Methane comes from natural gas stored in the Earth, or from methane deposits stored in a frozen form within the deep ocean sediments and in the Arctic permafrost. Rotting plant and animal matter release most atmospheric methane. Methane decreases as the air gets colder and is virtually absent by 120 km in the upper atmosphere.

- Nitrous oxide mainly comes from bacteria breaking down nitrogen compounds, but some is also formed by human industrial processes, such as the combustion of oil-based fuels. It decreases as the air gets colder and is virtually absent by 40 km.

- Ozone is a rare form of oxygen formed by UV radiation. It is very unstable and, while small amounts do reside in the troposphere between the Earth's surface and 20 km, relatively higher amounts exist in the stratosphere (20–60 km) and ozone is absent above 120 km. Ozone is a special gas when it comes to controlling the absorption and re-emission of radiation as it plays two roles: firstly, as a greenhouse gas in the troposphere; and secondly and most critically in the stratosphere, where it forms a protective blanket around the Earth by absorbing UV radiation from the Sun. Chlorine-related gases, expelled by volcanoes and chlorofluorocarbons (CFCs) formed by some industrial processes making aerosols, coolants and foams, can drift into the stratosphere and break down ozone, thereby increasing UV radiation towards the Earth's surface. Paradoxically ozone is also broken down by the very UV radiation responsible for its formation.

All of these gases interact with the world's oceans. If more gases are added to the atmosphere, then some of this gas is dissolved in the ocean. And, if the ocean temperature rises, some of the dissolved gases will be released back into the atmosphere. These processes of gaseous interchange are described by a law of physics known as Henry's Law.

A picky bunch

Wavelengths of longwave radiation energy are measured in microns and there are one million microns in one metre. Each greenhouse gas is very selective when absorbing longwave energy reflected from the Earth. The bandwidths of longwave radiation absorbed by various greenhouse gases are shown in Diagram 9.1:

- Water vapour: between 1 and 3 microns, 5 and 7 microns and then from 11 microns upwards.
- Carbon dioxide: around 2.5, 4.3 microns, and 11 microns, especially 14 to 16.5 microns.
- Methane: around 3.3 and 7.5 microns.
- Nitrous Oxide: around 4.8 and 8 microns.
- Ozone around 9.5 microns.

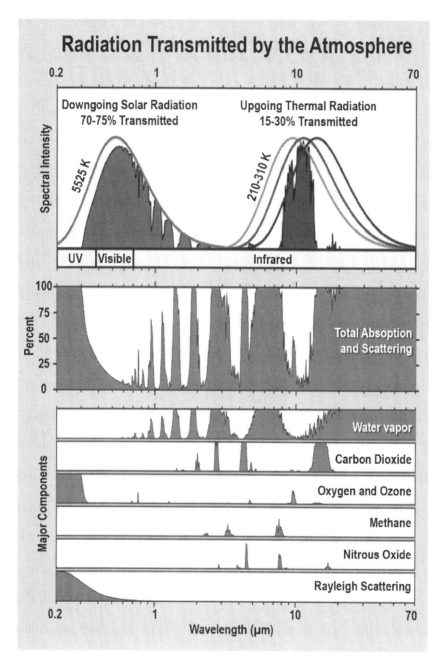

Diagram 9.1: The incoming and outgoing radiation, and the energy absorption behaviour of the greenhouse gases.

(http://commons.wikimedia.org/wiki/File:Atmospheric_Transmission.png)

This diagram clearly shows the different bands of radiation absorbed by the various gases.

1. *The top graph shows radiation coming from the Sun and radiation sent out from the Earth.*

2. *The second graph shows the total radiation absorbed by greenhouse gases. Note that some wavelengths are never absorbed.*

3. *The third graph shows the importance of water vapour – by far the main greenhouse gas.*

4. *The fourth graph shows the wavelengths absorbed by carbon dioxide.*

5. *The fifth graph shows the wavelengths absorbed by oxygen and ozone (a form of oxygen).*

6. *The sixth graph shows the FEW wavelengths absorbed by methane.*

7. *The seventh graph shows the FEW wavelengths absorbed by nitrous oxide.*

8. *The bottom graph shows the scattering of incoming light energy due to collisions with atoms in the atmosphere; this Rayleigh scattering causes the 'blue sky' effect.*

The diagram shows how greenhouse gases compete against each other. The importance of water vapour is very clear. More experimental data need to be collected on the emission of longwave radiation by different surfaces and on the absorption of longwave radiation at different latitudes and in diverse environments. Physicists are still relying on general physical theory about how surfaces and greenhouse gases should behave. They also rely on the concept of a *well mixed* atmosphere where gases disperse equally in all directions.

This is a pretty picky bunch! Each gas is like an animal that can only have a certain diet, and once it has had its fill, it cannot eat the leftovers. There is only so much each can do and they often compete with each other for the same food as there is only so much radiation each can absorb. This is a saturation effect. So after a greenhouse gas absorbs what it can of its favourite wavelengths, any extra radiant energy at that wavelength simply dissipates into space.

The atmosphere (a compressible gas) is naturally denser towards the Earth's surface where there are more greenhouse gas molecules per cubic metre. If more greenhouse gas is added to the atmosphere, more reflected radiant energy is absorbed closer to the Earth's surface. But, that reflected radiant energy is still finite as it is limited by the amount of energy sent out from the Sun and then reflected by the Earth's surface.

The order of merit

While various gases contribute to the greenhouse effect, *water vapour is the greenhouse king, many times more important than other greenhouse gases.* As discussed in the previous section, it absorbs over the greatest range of wavelengths. Due to the changing levels of water vapour and the fact that water vapour competes with other greenhouse gases, the relative role of the other greenhouse gas varies from place to place.

Emeritus Professor Antero Ollila of the Civil and Environmental Engineering Department of Aalto University, Finland has an excellent internet site with PowerPoint presentations on greenhouse gases and climate sensitivity (www.climatexam.com). The following table is only an average calculation of the role of each greenhouse gas, as in each area of the globe those calculations will vary (Ollila, 2014):

- Water 82.2%.
- Carbon Dioxide 11%.
- Ozone 5.2%.
- Methane 0.8%.
- Nitrous Oxide 0.8%.

The above calculations show the relative warming effect of each greenhouse gas, but they do not include other types of heat exchanges that occur such as when water vapour turns into cloud (or vice versa). They also do not include heat losses when radiation of the Sun is reflected straight back into space by cloud (an albedo effect), or heat insulation when clouds come over late in the day and act as a blanket giving a warmer night (an insulation effect).

Even if the other greenhouse gases were totally absent from the Earth's atmosphere, the Earth would be kept warm due to the greenhouse effect of water vapour. However, across the Earth, the water vapour content varies continually with the temperature and the pressure of the local atmosphere. Around the equatorial ocean the water content in the atmosphere is the highest and the greenhouse contributions of water vapour is estimated to be well over 90%, with that of carbon dioxide around 6%. Over the Antarctic polar plateau, where the surface temperatures are very low, there is almost no water vapour in the air, and the role of carbon dioxide is much higher.

Water vapour is so difficult to measure due to the enormous variations between arid areas with little atmospheric moisture (hot deserts or frigid polar regions) and tropical regions with high atmospheric water vapour.

Indeed, the tropics are not the hottest places on Earth due to added cooling from the evaporation of water. It is the deserts, where there is little water vapour, that are the hottest places on Earth.

The slippery eel

So, while water vapour is the *greenhouse king* in terms of warming the Earth, it is also the *slippery eel* as it influences the Earth's climate in confusing ways that could counteract or amplify or diminish its effect as a greenhouse gas.

The modelling of water vapour as the main player in global warming debate is a nightmare. How does one model the heat exchanges taking place when water is dumped by dew at dawn OR evaporated as the Sun rises OR when it condenses to cloud and mist OR when it leaves the atmosphere as rain or snow, and so forth? And how does one model different forms of cloud, at so many different altitudes? A cumulus cloud has an average life just over an hour, a cumulonimbus cloud around two hours and a nimbostratus cloud about three hours. Most clouds are minute water droplets, but the higher cirrus clouds consist of microscopic ice crystals. Time lapse photography shows the constant state of flux in this incredible dynamic world of cloud formation.

An empirical approach to document measurements of atmospheric water vapour, atmospheric temperatures and pressures is to gather data from radiosonde balloons. These balloons have battery-powered instruments that radio back data such as altitude, pressure, temperature, relative humidity, wind speed, direction, and even cosmic ray intensity. A typical radiosonde balloon reaches altitudes around 20 km. This height is near the top of the troposphere in the tropics and well into the stratosphere at high latitudes. Once thousands of data points are collected over a period of time, these balloons can build an empirical picture of what is happening.

The lowest zone in the atmosphere is called the troposphere, which is 20 km in height in the tropics and thins to around seven km at the poles. While most of the greenhouse effect takes place in the lower two kilometres of the troposphere any water vapour in the middle to high troposphere is a critical backstop because of the wider range of longwave energy it absorbs. The modelling of such water vapour is central to the climate debate. Most current climate models increase such high altitude water vapour as the world warms, but due to the nonlinear nature of the complex forces that determine climate, such an argument is simplistic.

The climate models add extra water vapour to the middle and upper troposphere especially in the tropics, but many atmospheric physicists say that water vapour is just not there.

Problems about the roles of water vapour and clouds in climate models will be discussed in the next chapter on climate modelling. It is clear that the IPCC position on global warming is absolutely dependent on calculations about the net warming effect of water vapour and clouds. *Should these calculations be wrong, the whole position of the IPCC collapses because the present climate debate is, to a large extent, about the role of water vapour in the global climate system.*

The second lieutenant

The second most important greenhouse gas is carbon dioxide and this is the greenhouse gas that has increased by 40% in the past 150 years.

In July 2014, NASA launched its Carbon Observatory satellite. This will provide a high resolution picture of carbon dioxide levels over all the Earth. The satellite can even be tilted to aim at a precise location for several hours. The first data indicate strong carbon dioxide plumes over some areas of the ocean (outgassing carbon dioxide), some carbon dioxide plumes over rainforest areas (due to land clearing and fires), and the only strong plumes of industrial carbon dioxide over China. Data from this satellite may change our knowledge of the carbon dioxide cycle, and provide data on the relative contribution of both man and nature to increasing carbon dioxide levels.

Physicists tell us that the warming effect of carbon dioxide lessens with each increase. This is because the warming effect of each doubling is the same. *So a 10 to 20 parts per million increase in carbon dioxide levels has the same warming effect as when these levels increase from 100 to 200 parts per million.* Consequently, at the start of the Industrial Revolution in Europe, a percentage increase in carbon dioxide levels at that time was more significant in effect than a percentage increase at the present time. This may surprise, but for example, the first 20 parts per million of carbon dioxide are calculated to have a warming effect of 15.3 watts per sq metre, while a 20 parts per million increase of carbon dioxide in the atmosphere at the present time only accounts for an extra 0.2 watts per sq metre!

Consequently, it is difficult to understand how some scientists attribute global warming in the past 50 years to rising carbon dioxide levels, and earlier warming in the 19th and early 20th centuries to natural forces. Even in the 19th century, carbon dioxide levels were rising and according to the physics of carbon dioxide a 1 part per million rise in carbon dioxide in the 19th century has a greater greenhouse effect than 1 part per million in the 21st century. Mathematicians describe this behaviour as *logarithmic*, so a one part per million increase when carbon dioxide levels were 280 part per million in the 19th century has the greenhouse effect of an increase of 1.4 parts per million when carbon dioxide levels reached 400 parts per million in the first decade of 21st century!

There is little debate among scientists about the physics of carbon dioxide as a greenhouse gas. *The present debate about future global warming is not about the warming effect of increasing carbon dioxide levels but about the role of clouds and water vapour in the atmosphere.* After all, in the climate models used by the IPCC, water vapour and cloud nearly triple the greenhouse effect of carbon dioxide. *If this amplifying effect is an over-calculation then the temperature and sea level scenarios most frequently touted in the IPCC documents are simply wrong.*

A stark conclusion

The various roles of greenhouse gases in determining our global climate involve complex multi-tasking. The idea that adding more greenhouse gases to the atmosphere will lead to a hotter and hotter world is too simplistic. Water vapour is the main greenhouse gas and water vapour can act in other complex ways with opposite effect and cool the atmosphere. As the story unfolds in the next chapter, there are serious problems with warming forecasts in IPCC documents based on what is called the **equilibrium climate sensitivity index** (a temperature rise forecast based on the doubling of carbon dioxide levels). Indeed, there is empirical evidence that the equilibrium climate sensitivity index used by the IPCC for the past 30 years is far too high. If so, the fiery greenhouse Armageddon predicted by the models may morph into a period of further benign, beneficial warming.

Chapter 10.

The models – we expected too much!

Model making

Billions of dollars have been spent creating computer models to mimic climate change. These models are so complex that not all are built from scratch. Like model cars, different computer models may share some of the same blocks of code. If an institution wants to set up its own model, there can be a lot of selective horse-trading with other institutions to swap desirable blocks of computer code. This means that an institution may not completely understand all the code in its own model, and may even find it difficult to revise some sections of code it did not develop.

Since climate occurs in three dimensional space, one model is made for the Earth's atmosphere to a particular altitude and another for the ocean to a certain depth. Each of these models has a basic cell that is so many kilometres long and wide, and so many metres deep. The two models are then integrated to create what is called a General Circulation Model or GCM. It is like rooms in a skyscraper exchanging heat with the rooms above and below and then with the rooms in the skyscrapers next door.

There has to be a trade-off with the computer cell size. If the cell size is too small even powerful computers would take years to run these GCMs. Indeed, a quirky mathematician can point out that *if the cell size was reduced to cubic metres the time to run the computer program would exceed the age of the universe!*

Despite the power of modern computers the cell sizes are still large. In the model used by British Hadley Centre the ocean cell is 1540 cubic kilometres and the atmospheric cell is 29,000 cubic kilometres. Diagram 10.1 shows the latest improved model at the British Hadley Centre. This latest model uses grid blocks at the equator as follows:

- Atmospheric model: the blocks measure 1.875 degrees of longitude by 1.25 degrees of latitude and one km in depth, and there are 38 levels. At the equator each grid block is 29,061 sq km (209 km by 139 km).

- Ocean model: the blocks measure one degree of longitude by one degree of latitude and 125 metres in depth, and, depending on the depth of the ocean there are 40 levels. At the equator the blocks are 12,321 sq km (111 km by 111 km).

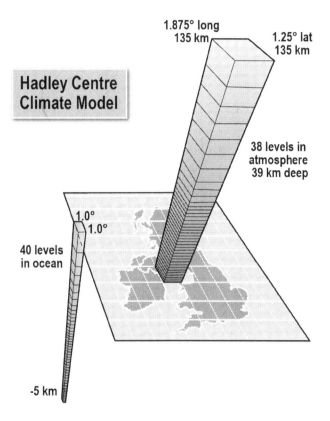

Diagram 10.1: The climate model used by the British Hadley Centre, England (www.metoffice.gov.uk).

The limitations of the model are clear once one realises that only one figure can be placed in each cell for a particular feature, such as temperature. A critical problem is that as the grid size gets smaller, the computer time increases. If the grid blocks are too small, the computer time to run the model could be millions of years.

Model headaches

The computer models simply inform us how the climate data put into the model *(the knowns)* interact. There may be critical inputs that are not in the models *(the unknowns)*, but, even then, we cannot infer if the *unknowns* were inserted into the model, everything would be okay and the model would then work. That assumption is also false. There is still the inherent quirkiness of nonlinear/chaotic systems, so that any idea of a *eureka* method that will predict climate over aeons is absurd.

Even though the IPCC has admitted its models are not predictive tools, it still treats the future scenarios in the climate models as reasonable predictions. *But, what good is a model doing something that even the IPCC says it cannot do?* If an outcome is not possible, it cannot be said to be a reasonable prediction. This is a central contradiction even commented on by the IPCC:

> *In climate research and modelling, we should recognise that we are dealing with a coupled nonlinear chaotic system, and therefore long-term prediction of future climate states is not possible* (IPCC 2001 report, Section 14.2.2.2).

The founder of nonlinear/chaos theory, Edward Lorenz, commented 40 years ago to the effect, that even if we had information for every cubic metre of the atmosphere and ocean and built the most wonderful supercomputers, we still could not forecast the weather in a month's time. Lorenz realised the problem of nonlinearity on a day he was in a hurry. In order to run his weather program he entered 0.506 from a printout instead of entering the full precision value of 0.506127. Then when his model was run it delivered astoundingly different results. This is known as the *butterfly effect*: musings that the flapping of the wings of a butterfly some weeks earlier may have some influence on the outcome of a hurricane. The development of chaos theory, and how minute changes in initial conditions may even change faraway events, is well described in *Chaos – making a new science,* the famous book on nonlinear theory by James Glieck (Glieck, 1988). The book is a classic and earned Glieck, a Pulitzer Prize.

Tuning

To get the models to work, best estimates or what are called *parameters* for various complex factors have to be inserted into the models, and even then these estimates are varied while the models are running. This is called tuning. The insertion of parameters into the models enables the models to approximate what is happening in the real world. There is a certain amount of circular reasoning in this tuning. So a model is not necessarily correct just because it predicts how phenomena are behaving. That correctness may just be the result of tuning.

The tuning of scientific models is not new; the Greeks were at that game over 2000 years ago. For example, because the planets cover equal areas of their elliptical orbits in equal times, their speed is slower at the long ends of their orbits and faster in the narrow sections of their orbits. This differential speed means that at certain times the planet Earth is travelling faster than say Mars, so Mars seems to be going backward if it is plotted in the sky at the same time on successive nights. The Greeks knew this and Eudoxus of Cnidus in the 4th century BC attached all the planets to giant glass spheres rotating at different angles and speeds. His model had 27 glass spheres, but then his student Callippus of Cyzicus found anomalies and so he added seven more spheres to better TUNE the model.

One hundred years later, the Greek expert on the mathematics of conics, Apollonius of Tyana, came up with a new model where each planet was attached to a small wheel attached to the rim of a larger wheel, and the larger wheel went around the Earth. This meant that sometimes a planet, moving backwards on the smaller wheel, could appear to retrograde in the sky at certain times. A further 300 years later, in the second century AD Claudius Ptolemy, who was living in the Roman Province of Egypt, took the ideas of Apollonius and devised a more complex system of *epicycles* and his system was used for the next 13 centuries. However, over those centuries, as more anomalies were found, the Ptolemaic system to explain planetary and stellar motions was further TUNED! By the 13th century AD there were 40–60 large wheels and 40–60 small epicycles.

A most amusing model was created by two Argentinian mathematicians, Christian Carman and Ramiro Serra. In 2005 they created a system of 10,000 epicycles that traced the cartoon figure of Homer Simpson. This was a quirky way to show that models could be TUNED to fit the weirdest configurations (Bellos, 2014).

So back to computer models. In order to run these models programmers insert their best observations and calculations for how much rain and snow is precipitated, for relative humidity values, and for the extent of cloud cover, and so on. The computer programmers acknowledge that some parameters are inserted to represent unresolved physical processes. But, if these processes are unresolved, how much faith can we have in the models that, to a great extent, depend on these parameters for their final outcomes?

Going separate ways

In 1975 NASA began to collect world wide temperature data using satellites. NASA satellites use microwaves to measure the temperature in the lower atmosphere up to an altitude of eight kilometres. The project is managed for NASA by the University of Alabama, Huntsville. Diagram 10.2 shows the temperature record since 1983.

Between 1975 and 1998 temperatures and carbon dioxide levels were both rising in tandem. Then, after 1998 the temperatures predicted by the computer models and the actual world temperatures measured by NASA's satellites began to diverge. Indeed, between 1998 and 2017, carbon dioxide levels rose over 9% while temperatures stayed level, a trend against all model predictions. This pause has lasted 19 years and is still ongoing in 2017.

As this divergence between model-land predictions and real data increased, so did the embarrassment for climate science. The models developed over the past 40 years were clearly wrong. Yet, this embarrassment should have been expected. Historical records (mentioned previously in Chapter 4) show that there were other recent periods when carbon dioxide levels rose and temperatures moved sideways. Congruence of two variables over time or throughout space is not always an explanation of their relationship.

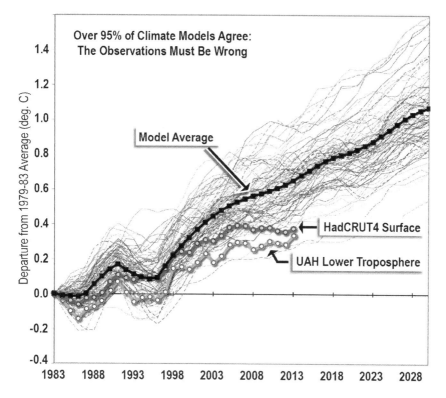

Diagram 10.2: NASA satellite temperature measurements 1983–2013 (UAH), the British Hadley Centre surface temperature chart for the world 1983–2013 (HadCRUT4 Surface) and the temperature projections of 90 climate models (from the website of Professor R. Spencer – www.drroyspencer.com).

It is clear that temperatures have paused after 2000 and are going sideways. There is no coherent trend between temperature and greenhouse levels and the models since 1995. The satellite data are reported as UAH (University of Alabama, Huntsville from base year 1979) and the data come from a number of satellites measuring radiation at various wavelengths, from which air temperatures are calculated.

It is most comical to read one of the emails that came to light when the University of East Anglia's computer system was hacked in November 2009. A Dr Kevin Trenberth, a leading IPCC author from the National Centre for Atmospheric Research in Boulder, Colorado, is confronted with satellite data showing that the Earth's temperature has paused while the climate models show warming. He comments:

The fact is that we can't account for the lack of warming at the moment and it is a travesty that we can't. The CERES (Clouds Earth Radiant Energy System) data published in the August BAMS (Bureau of the American Metrological Society) 09 supplement on 2008 show there should be even more warming: but the data are surely wrong. Our observing system is inadequate.

Dr Trenberth suspected that the data must be wrong, that the models must be right, and that if we looked more closely we would find data that are consistent with the models. Unfortunately for Dr Trenberth, the temperature pause he thought was an artefact of wrong data in 2008 is still in place in 2017, nine years later.

A stubborn riposte

A common response of the scientific community is that the models have *not failed*.

Many scientists maintain that the world is still warming and that the heat predicted by the models is not in the atmosphere, but is *buried* in the ocean. However, Dr Xinfeng Liang, from the Massachusetts Institute of Technology (MIT), Professor Carl Wunsch from Harvard University and other co-authors from MIT, counter this argument. *They show that recent oceanic heat uptake from the atmosphere is very small, in the order of a few tenths of a watt per square metre; too small to affect the deeper ocean.* They also contend that some areas of the ocean are warming through heat coming from the ocean bodies below them, and that other parts of the abyssal ocean are cooling. It is clear from their work that *heat exchanges* in each of the world's oceans, although extremely complex, are the *result of energy exchanges over many years*, and not the result of global warming in the past few decades (Liang et al., 2015).

Other scientists use a different argument to defend their climate models. They contend that the models have not picked up the cooling effect of man-made aerosols, such as dust and industrial pollutants or gases from volcanic eruptions. They argue these aerosols create high-altitude haze that blocks the Sun's radiation, and so cause pauses in global warming. For these scientists the aerosols are simply masking the heating calculated by the models. However, these man-made aerosols are not produced in a cyclical fashion. And how can they explain a pause late in the 19th century when industrial pollution was much lower? And why aren't these long temperature pauses in the heavy industrialised Northern Hemisphere radically different from those in the more lightly industrialised Southern Hemisphere? And why are these pauses so regular in character?

There is also no cyclical pattern in volcanic eruptions to explain regular warm/pause cycles. Cooling from volcanic events can be widespread only when very violent and large eruptions eject large quantities of dust into the stratosphere. In recent times the largest eruption was the 1815 Tambora eruption. It occurred during the cold Dalton Minimum event that began in 1795 and its cooling effects were only noticeable for a few years.

Climate sensitivity

It is now becoming clear that there was a *ticking time bomb* in the computer climate models that had been developed between 1975 and 1998 period. This ticking time bomb was an index called the Equilibrium Climate Sensitivity index (ECS); an estimate of the temperature change if carbon dioxide levels doubled. In 1979, the American Academy of Science met for one week and determined the Equilibrium Climate Sensitivity index to lie between 2°C and 4°C with an average of 3°C. These estimates are still used by the IPCC. The variables considered in developing this sensitivity index were:

· The doubling of carbon dioxide.

· The subsequent increase in water vapour, as temperatures rises.

· The decrease in snow and ice cover (there is a presumption that as the surface of the Earth warms there will be more exposed areas of soil and rock, which in turn absorb more heat and increase the warming effect).

· Estimates of cloud cover.

Because of the large number of complex variables, this climate sensitivity index is *estimated* by physicists. It not derived from one simple physics equation, so there is a large range of estimates.

The ECS index used by the IPCC for the past 30 years has an average value of 3.0°C. That index has been used in all the IPCC reports. Within that 3.0°C, the warming effect of carbon dioxide is estimated to be 1.0°C, the net warming effect of water vapour is 0.7°C, and the net warming effect of clouds is 1.3°C. So, as carbon dioxide levels double, water vapour and clouds nearly triple the effect of carbon dioxide. *This means that the climate debate is better described as a debate about water vapour and clouds, as they are more critical than carbon dioxide. If they are calculated wrongly, then climate model-land crashes.*

It is not hard to see that if water vapour and clouds increase the climate sensitivity index to 3.0°C, the computer model-land programmes will lead to doomsday thinking with predictions of sea levels and temperatures accelerating and polar ice sheets collapsing.

However, even though sea level is rising, as explained in Chapter 6, there has been no long-term acceleration of sea level in the past 150 years. This fact alone casts grave doubt on high estimates for the climate sensitivity index. It is also hard to see how there could have been any periods when global temperatures move sideways while carbon dioxide levels rose if the climate sensitivity index was so high.

Many scientists have pointed out serious problems with the computer modelling of climate. Here some of these problems are discussed; enough to show that the science of global warming is far from settled.

Water vapour headaches

The relative humidity conundrum

The climate models used by the IPCC presume that rising temperatures produce more evaporation and therefore drive more water vapour into the atmosphere. This argument looks very logical. In Chapter 5, called the Weather Armageddon, it was shown that many have wrongly argued that rising temperatures mean more rain, more wind and more violent storms. Yet, the actual data recorded at meteorological stations worldwide show no change in the frequency of severe weather in the 20th century.

It seems that the argument about water vapour and its warming effects also used overly simplistic logic. Relative humidity is the amount of water in a given volume of air compared with the maximum amount of water vapour that same volume of air contains at a given temperature and pressure. If the atmosphere heated up and the same amount of water vapour stayed in the air, relative humidity would drop because now that same volume of air can hold more water vapour. Presuming that water vapour will rise, as temperature rises, the computer models are programmed so that as temperatures rises, the amount of water vapour increases and so relative humidity stays around the same levels.

But, are conditions in nature as simple as this? Dr Richard Lindzen points out that when we walk around on a humid day, the atmospheric humidity could increase and slow down the evaporation of water. If, for example, the relative humidity rose from 80% to 85%, that rise would choke back evaporation and the evaporation would not be more than when temperatures were 3°C lower and the relative humidity 80% (Lindzen in Moran, 2014).

Troublesome wind

This is another complicating argument. The amount of water vapour in the air is not only dependent on temperature but also on wind. There is no comparison between the rate of evaporation when the air is still and the temperature is around 30°C, with the situation where the temperature is only 10°C, but a strong wind is blowing. There is much more evaporation in the much colder conditions due to high wind. Wind can be far more critical to the amount of water vapour in the air than temperature. So if global warming means there is a lower temperature gradient between the tropics and the lower latitudes, can this mean a drop in average wind velocities in mid-latitudes? This would mean linking rising water vapour with rising temperatures in the climate models is too simplistic.

A story illustrating the dramatic example of the evaporative power of high wind acted out during a few Antarctic summers in the 1970s. In 1969 a group of Japanese scientists found nine meteorite fragments in blue ice at the foot of the Yamato Mountains in Antarctica. They had been spurred on by a throw-away conjecture by a Japanese scientist who had never ventured into the polar regions. He theorised that high winds falling from the cold Antarctic plateau could evaporate glacial ice and expose meteorites. Basically, if the ice sheet is kilometres thick and there is a rock on the surface nowhere near mountains, it could come from outer space. The Japanese scientist correctly pointed out that the smooth, blue ice areas on the glaciers dropping down from the polar plateau could expose a treasure trove of meteorites. He was so right. Even though wind velocities are low over the large continental interior of Antarctica, this huge body of cold air plummets from the edge of the plateau at 3000 metres to sea level all around Antarctica. This means the 300 kilometre edge of Antarctica is the windiest place on earth.

In 1969, the Japanese found a few meteorites near their Antarctic Showa Station, but then during the 1970s the Americans, in true American style, found so many meteorites poking out of the blue glacial ice on the edge of the polar plateau that their number exceeded the collection of meteorites in world museums at that time. And, to complete this story, they also found rocks in the ice that were blasted off the moon when it was bombarded by meteorites. *So the amount of moon rock found in Antarctica has exceeded in weight the moon rocks already brought back by the lunar astronauts at huge cost.* At the present time, the Americans have collected over 20,000 meteorite fragments from Antarctica. Yet all this work uncovering meteorites and evaporating huge quantities of ice was done by wind at temperatures less than –30°; such is the power of wind.

Another climate scientist who has questioned assumptions of a constant relative humidity, especially in the atmosphere above five kilometres, is Garth Paltridge, an Emeritus Professor of the University of Tasmania and a former chief research atmospheric scientist of the Commonwealth Scientific and Industrial Research Organisation of Australia (CSIRO). *He showed a declining trend in relative humidity in the atmosphere above 3 km in the 40 years since 1970.* The data came from thousands of radiosonde balloons. The data are contrary to assumptions in the IPCC models that maintain a constant relative humidity and have higher water vapour levels at these altitudes. The data show, according to Paltridge, that relative humidity has only risen in recent years in the tropics and southern hemisphere at altitudes less than 1 km. This would seem to imply frequent changes in relative humidity and a drying upper atmosphere, not in line with computer models used by the IPCC. Paltridge in his book *The Climate Caper*, recounts the difficulty of publishing a scientific paper about this problem that was based on data, not on abstract theory (Paltridge, 2009).

The freeze-drying clouds

There is a further argument that may explain the dry middle troposphere described by Paltridge. Dr Bill Gray (now deceased) was an American atmospheric physicist and an Emeritus Professor from Colorado State University who specialised in hurricane research throughout his scientific career. In his view, *tropical cumulo-nimbus clouds freeze-dry the middle troposphere.* Dr Gray contends that these clouds, when rising, overshoot the position in the atmosphere that matches their temperature and water content. Like a train without brakes these clouds rush higher into even colder air, and this means they are then more efficient at releasing their water content as rain or snow. So in this theory, such high altitude clouds work in a nonlinear fashion and rather than increasing water vapour in the upper atmosphere actually decrease it. This anomaly means there is a marked difference between a moist lower troposphere and a drying upper troposphere. For Dr Gray these cumulo-nimbus clouds should be treated as a negative cooling factor in any climate sensitivity index; the opposite to their present treatment in the climate models. This scenario is not inconsistent with the situation described by Paltridge.

The missing hot spot

Finally, water vapour in the climate models has a positive warming effect. The models used by the IPCC predict a theoretical tropical *hot spot* in the mid-troposphere (around an altitude of 10–15 km). This is due to the heat given out when water vapour condenses. Dramatic diagrams of this hot spot were in the earlier IPCC reports. *Yet, this theoretical hot spot has never been proven, despite over 20 million radiosonde balloon readings taken in tropical regions.* While these diagrams are now absent in the IPCC reports, the IPCC models still predict that the tropical hot spot exists.

Dr Steven Japar was involved in the 1995 and 2001 IPCC reports and he resigned over the absence of this hot spot anomaly. Dr Japar is an atmospheric physicist who has published over 80 peer reviewed scientific papers on climate change, atmospheric chemistry, air pollution and vehicle emissions. He comments:

> *Temperature measurements show that the hot zone, that is predicted by the models in the mid-troposphere, is non-existent. This is more than sufficient to invalidate global climate models and projections made by them.*

It gets worse

Starting date assumptions

The climate models need to start at a particular date with a standard atmosphere that changes over time as variables are altered. The common starting date for climate models is 1750 AD. When modelling started, the only estimate for a standard atmosphere had been done in the United States. This meant that the average US atmosphere in 1976 was used as the standard starting *'global atmosphere'* even though *it is very debatable that this atmosphere, called US76, equates to the average world atmosphere.* Indeed, US76 holds about 7500 parts per million water. This gives the models an initial water vapour assumption that is now considered to be much lower than the world *Average General Atmosphere* (AGA). According to Ollila the average world general atmosphere holds around twice the water vapour content of US76, and the model's use of an initial low starting point wrongly increased the sensitivity of the climate models to increases in water vapour (Ollila, 2015).

Heat transfer headaches

There are problems with the mathematical formulae in the models that try to handle the transfer of heat in turbulent conditions in the atmosphere and in fluids; yet the transfer of heat in the ocean and in the atmosphere often occurs in turbulent conditions. *It is not proven that such equations (the Navier-Stokes equations) always work in three dimensions or just have one solution.*

Indeed, the models have trouble handling the transfer of heat in the atmosphere from the equator to the middle and polar latitudes along what is called is called the EPTG (the equator to pole temperature gradient). The models also have trouble handling the heat exchanges between the ocean and the atmosphere during well known cycles, such as the El Nino–Southern Oscillation cycle (ENSO) cycle, the Pacific Decadal Oscillation (PDO) cycle or the Atlantic Meridional Oscillation (AMO) cycle. The IPCC models also do not address heat exchanges between the upper troposphere and lower stratosphere due to ozone formation caused by the Sun's output of UV rays – an output that varies as much as 75% during each solar cycle.

With regard to heat exchanges in the ocean, it must be noted that the heat capacity of the upper few metres of the ocean equals the total heat capacity of the atmosphere at any time and our knowledge of the heat and temperature profiles of the ocean is extremely limited. Professor Judith Curry of Georgia Institute of Technology and Dr Nicholas Lewis from Bath, England comment:

> The ocean accounts for over 90% of total estimated energy accumulation and for almost all uncertainty (Curry and Lewis, 2014).

Albedo presumptions

Albedo refers to the ability of a surface to reflect heat back into space. Black bodies have low albedo and absorb heat, while white surfaces, such as cloud, snow and ice, have a high albedo. There is a dramatic black body effect when you place a slab of dark chocolate on an icy surface; that slab of chocolate will absorb heat and melt its way into the ice!

The models presume that as the Earth warms the total area under ice and snow decreases and the albedo of the Earth's surface decreases. The largest decrease in the Earth's albedo in recent years has been in the Arctic as the summer Arctic pack ice has declined in area by about three million sq km. But the Antarctic pack ice has defiantly increased each year for the past 40 years, and more than cancelled out any decrease in the cover of Arctic pack ice at the North Pole (see Chapter 8). Therefore, in recent years there may have been no significant change in the albedo effect.

Aerosol tuning

One forcing factor used to tune the climate models is the cooling effect of man-made aerosols (smoke, dust, industrial pollutants). These can reflect solar radiation back into space (back-scattering) and have a cooling effect. In 2014, Lindzen pointed out that the IPCC models are not consistent as they use different *aerosol* adjustments for all different and higher estimates of ECS (Lindzen in Moran, 2014).

More on clouds

There are scientists who estimate that clouds have a net negative cooling effect. One of these is Professor Roy Warren Spencer, Principal Research Scientist at the University of Alabama in Huntsville, who for the past 20 years has been one of the managers and principal investigators of the NASA satellites measuring global temperatures in the lower atmosphere (Spencer and Braswell, 2011). Another is Dr Richard Lindzen, one of the world's leading atmospheric physicists, who argues that upper level cirrus clouds cancel any positive water vapour feedback (Lindzen in Moran, 2010).

Dr Jonathan Gero and Dr David Turner from the University of Wisconsin Space Science and Engineering Centre have analysed 800,000 measurements of long wave infra-red radiation over the US southern plains. This data set spans 14 years. The down-welling radiation decreased in winter, summer and autumn and increased only in spring. The data show more than expected outgoing long wave radiation escaping to space in contradiction to the predictions of standard climate models. The trends were statistically significant and were also related to cloudiness. So for most of the year there was a drier upper atmosphere than suggested by the climate models and that drier atmosphere led to increased loss of heat into space (Gero and Turner, 2011).

Aerosols can have another critical function. Their fine particles can become the nuclei of cloud droplets as moisture condenses on them. And since dust and industrial pollution are mainly at lower altitudes, such aerosols can increase low cloud formation. These low clouds can reflect solar radiation back into space and have a cooling effect during the day but, if they form late in the day, heat radiated from the Earth's surface is kept within the lower atmosphere during the night. With such diverse functions, such clouds are a source of uncertainty in the climate models.

Since clouds can reflect sunlight back into space or reflect back radiation from the Earth or generate evaporative cooling as they dissipate, *clouds are listed as a major uncertainty in all climate change models.* Indeed, cloud cover cannot be seen in any detail at the resolution of the models due to inadequate cell size. This means that crude estimations of cloud cover and of the thickness of clouds at varying altitudes have to be inserted into the models.

There are many statements in the IPCC reports that underscore the uncertainty in scientific reports of the roles clouds play in the global climate system. An excellent technical summary of cloud feedbacks is in the IPCC Fourth Assessment Report: Climate Change: Section 8.6.3.2: 2007:

> *In many climate models, details in the representation of clouds can substantially affect the model estimates of cloud feedback and climate sensitivity (Senior and Mitchell, 1993; Le Treut et al., 1994; Yao and Del Genio, 2002; Zhang, 2004; Stainforth et al., 2005; Yokohata et al., 2005). Moreover, the spread of climate sensitivity estimates among current models arises primarily from inter-model differences in cloud feedbacks (Colman, 2003a; Soden and Held, 2006; Webb et al., 2006; Section 8.6.2, Figure 8.14). Therefore, cloud feedbacks remain the largest source of uncertainty in climate sensitivity estimates...*

> *In doubled atmospheric CO_2 equilibrium experiments performed by mixed-layer ocean-atmosphere models as well as in transient climate change integrations performed by fully coupled ocean-atmosphere models, models exhibit a large range of global cloud feedbacks, with roughly half of the climate models predicting a more negative Cloud Radioactive Forcing (CRF) in response to global warming, and half predicting the opposite (Soden and Held, 2006; Webb et al., 2006).*

This comment by Professor Ian Plimer in his book *Heaven and Earth is so apt:*

> *A change of just 1% in the cloudiness of planet Earth could account for all the 20th century warming. However, the IPCC computers don't do clouds* (Plimer, 2009).

It is interesting to note that Professor Richard Lindzen, Professor of Meteorology at the Massachusetts Institute of Technology between 1983 and 2013, resigned from the IPCC because the Summary Statements in the IPCC reports did not correctly reflect what scientists were saying about clouds in the larger technical volumes.

A lower ECS

It is not surprising that some scientists have re-visited the Equilibrium Climate Sensitivity index (ECS) used in the climate models. Their estimates of the ECS are much lower. After all:

- Temperatures have risen less than 1.0°C while carbon dioxide levels have risen 40% since 1850. However, a 40% rise in carbon dioxide levels according to the physicists is equivalent to about 80% of the doubling effect of carbon dioxide taken from that date. This is because each increase in carbon dioxide has a lesser effect; a behaviour described as logarithmic by mathematicians.

- As noted in Chapter 6, sea level, although rising, is not accelerating.

- Temperatures have often paused while carbon dioxide levels increased indicating that the effect of rising carbon dioxide levels cannot be that dominant within the present, complex climate system.

Within the past five years, some ECS estimates have been 1/2 to 1/6 of the average value of 3.0°C used by the IPCC, such as,

- An average of 1.65°C (Lewis and Curry, 2014).

- Averages of 0.46°C, 0.56°C, and 0.58°C (estimating ECS using 3 different methods, Ollila, 2014).

- An average of 0.5°C° (Lindzen and Choi, 2009).

- An average of 0.482C° (Miskolczi, 2010).

The divergence between the models and real data has finally had some influence on the IPCC but it is only making some minor changes to its general position. In 2013 the IPCC did change the lower limit of the ECS from 2°C to 1.5°C but left 4°C as the upper limit and, strangely, did not publish a revised average.

In 1995, the IPCC predicted that world temperatures would rise 0.7°C between 1995 and 2025. But, since world temperatures paused after 1998, in September 2013 the IPCC revised its forecasted rise for this period downwards from 0.7°C to 0.5°C. This is an implicit acknowledgment that the ECS used by the IPCC is too high. Yet, even now this revision is inadequate. The amended IPCC forecast means there needs to be a temperature increase of 0.05°C a year between 2013 and 2025. Originally, the annual temperature increase predicted by the IPCC was 0.017°C per year, but now the IPCC's forecast requires a high catch-up rate of 0.05°C/year; a rate equivalent to three times the original rate, and equivalent to a temperature increase of 5°C/century; a high rate of warming not even envisioned in any IPCC long-term models.

No longer swamped

The high ECS used by the IPCC has virtually ruled out other natural forces as an explanation for recent temperature changes. If the ECS is much lower, it is far easier for other climate forces to over-ride the greenhouse effect and cause global temperatures to fluctuate. Some of these small climate forces are: small variations in solar energy due to changes in solar radius; the warming effects of ozone formation in the lower stratosphere; small variations in Total Solar Irradiance due to wobbles in the solar orbit; changes in cloudiness caused by the Sun's magnetic field changing the amount of cosmic rays reaching the Earth; heat transfers between the Earth's oceans and its atmosphere during oceanic cycles, such as the Pacific Decadal Oscillation (Diagram 10.3).

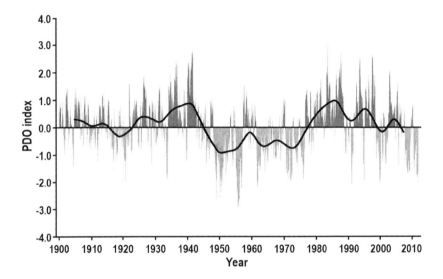

Diagram 10.3: The Pacific Decadal Oscillation (Diagram from the University of Washington Climate Impacts Group).

This diagram of the PDO show: the warming trend between 1915 and 1940; the cooling phase between 1945 and 75; the warming phase between 1975 and 1998; and a more complex downturn between 1998 and the present. While this PDO may be coupled with the warm/pause cyclicity of world temperatures over the past 150 years, it is difficult to see how it could, on its own, be a driving force for temperature on a global scale even if the climate sensitivity of the Earth's climate system to changing carbon dioxide levels was very low. However, a low climate sensitivity index (ECS) could allow variations in the input and output of heat energy from the world's oceans to have a more significant impact on global climate.

Previously, such smaller warming or cooling forces would have been swamped by the high warming estimates derived from the ECS formulae used by the IPCC that clearly led models to overestimate the warming effect of increasing carbon dioxide levels. The high ECS had blotted out the Sun. As the next chapter explains, a *lower ECS also brings Father Sun back onto the central stage of the climate debate.*

A respect for history

If earlier historical climatic patterns during the past 150 years had been recognised as valid predictive tools, physicists would have carefully examined earlier pauses in world temperature, not just the recent pause. Then, they would not have been assigned such importance to climate models using a high sensitivity index.

As Stephen Hawking put it so clearly in his famous book, A *Brief History of Time:*

> A theory is a good theory if it satisfies two requirements. It must accurately describe a large class of observations on the basis of a model that only contains a few arbitrary elements, and it must make definite predictions about the results of future observations (Hawking, 1996).

If the present climate models are tested and run from the start of the 20th century, they do not predict that 40% of the time global temperatures would remain static. Such a large divergence from what actually has happened is more than sufficient to invalidate the present climate models that have induced panic and led to costly, wasteful carbon reduction schemes and crazy economic decisions.

Historical patterns, even if not understood, at least contain the full nonlinearity of the system and need much more serious consideration. In this regard the words of Professor Jan Veizer, Emeritus Professor of Geology at the University of Ottawa, Canada are very apt:

> Models and empirical observations are both indispensable tools of science, yet when discrepancies arise, observations should carry more weight than theory (Veizer, 2005)

Chapter 11.

Is the sun still king?

White-washing the Sun

There has been so much focus on greenhouse gases as the primary cause of recent global warming that the Sun has been relegated to the background, somewhat like a steady light globe high in the sky, not doing much at all except giving out a steady light.

The term *unprecedented* is used with such abandon to describe increases in carbon dioxide levels, rising temperatures, increasing sea levels or higher storm frequencies. This tagging has placed our Earth in new territory and divorced the present Earth condition from its historical past. This artificial compartmentalisation has whitewashed the influence of the Sun and has divorced modern climate science from the past. For how can the past climate history be used as a reference map in the climate debate if we have entered such *uncharted, unprecedented waters*?

But is the Sun a steady light globe or is the Sun like an illusionist standing on an open stage in front of a captivated audience and daring them in a mocking voice, *"I may seem to be doing nothing, but I can make things change before your very eyes, and I challenge you to find out how I did it!"*?

Some puzzles to be solved by the Sun's dare are:
- How can the Sun's radiant energy change over short periods?
- Does the Sun's UV radiation influence climate on Earth?
- Can the magnetic field of the Sun influence climate?
- Can the Sun alter cosmic radiation from outer space reaching the Earth?
- How are sunspots and the Sun's magnetic field linked?
- How are low sunspot numbers linked to cold periods on Earth?

The star player

The Sun is our source of life, the foundation of our very existence, the underlying force behind our climate. The Sun is the king of the whole solar system. Sheer numbers give some idea of the size of this bright red ball that rises in the east each morning. Circumnavigation of the Earth involves a trip of 40,000 kilometres, but the same trip around the Sun would be 4.4 million kilometres. The Sun contains 99.8% of the solar system's mass and is 330,000 times the weight of Planet Earth and 1000 times the weight of Jupiter.

Due to the nuclear fusion of hydrogen, the Sun is a continuous fire-ball over 10,000,000 °C at its centre and around 6000°C in its 400 km thick outer layer, known as the photosphere. It is this outer layer that emits the Sun's radiant heat energy into space. The sun's electromagnetic radiation includes ultra short gamma rays, X-rays, UV rays, the visible light spectrum, the longer wave radiant heat spectrum and even radio waves. While the amount of radiant energy that reaches the outer atmosphere of the Earth is fairly steady, there are large variations within the Sun's shorter wavelengths that reach the Earth, such as UV rays and X-rays.

The Sun's nuclear explosions cast into space a stream of high energy particles that bombard the solar system. These fragments are a mix of sub-atomic particles such as electrons, neutrons, protons that make up what is known as *the solar wind.* This spray of energy comprises a billion, billion, billion, billion (yes that's right) particles per second! The Sun is actually losing the weight of the Earth every 150 million years!

The Sun also surrounds its solar system with an enormous magnetic field. This magnetic field is a three-dimensional force-field that can repel charged particles travelling at near the speed of light coming from distant exploding stars and galaxies. This magnetic field is so strong due to immense frictional forces between different zones of the Sun. This friction is between the inner two thirds of the Sun (rotating once every 30 days), the outer zone near the equator (rotating every 25 days) and the outer zone near the polar regions (rotating every 36 days). It is like a series of brake pads hitting discs spinning at supersonic speed. Such friction breaks up atoms and creates huge numbers of charged particles. These electric charges have created a complex magnetic field whose magnetic lines of force continually criss-cross and interfere with each other.

No escape

There are many different relationships between the Sun and the Earth:

- Sunspots on the Sun indicate changes in the sun's magnetic field which surrounds the Earth.
- Small wobbles in the Sun's orbit cause short-term variations in the distance between the Sun and Earth.
- Frequent variations in the Sun's output of X-rays and UV rays have profound effects on the outer atmosphere of the Earth and the formation of ozone in the Earth's atmosphere.
- Small changes in the Sun's radius and surface area vary the heat from the Sun that reaches Earth.

The spotted Sun

Dark spots on the Sun were noted by Chinese, Korean and Japanese astronomers over 2000 years ago, but they were not reported in Europe until the 17th century. These dark spots on the Sun are really like twinned worm-holes in the Sun's magnetic field. These sunspots look dark only because they are cooler than their surrounds as they are areas of down-welling gases. They appear in a cycle as the orientation of the Sun's turbulent magnetic field changes from a north-south polar orientation to an east-west equatorial orientation and then the field reverses its polarity. Each cycle lasts 11–13 years and is related to a switch in the Sun's magnetic field. At the start of a cycle there are few sunspots and they reach their maximum number in the middle of the cycle and then decrease again. At the crest of the cycle the sunspots are orientated in a rough east-west configuration between 25 and 35 degrees north of the Sun's equator and in the next cycle the sunspots are arranged in a similar configuration south of the Sun's equator.

Even though sunspots were first noted in the West in the 17th century, the numbering of the 11–13 year sunspot cycles only began when the 1755 to 1766 cycle was called cycle No.1. Our knowledge of the orientation of these sunspots comes from 400 years of observations. Professor Robert Baker and Professor Peter Flood from the University of New England, Australia, note it is possible that the orientation of these sunspots and the length of the cycles may have varied in the past as sunspots have been noted on other stars, such as AB Doradus and EK Draconis, where the sunspot configurations are different. In these stars the sunspot orientation at the crest of the cycles is north-south against the east-west configuration of sunspots at the crest of the sunspot cycles on the Sun (Baker & Flood, 2015).

Connections between sunspots and climate will be discussed later in this chapter. In the past, periods without sunspots certainly correspond with periods of colder climate. It is possible that the sunspot variations reflect small changes in the Sun's radiant energy output. It is also argued that the magnetic field of the Sun determines the pattern of the sunspot cycles and also controls, as a *gatekeeper*, the amount cosmic of radiation reaching the Earth – as we will soon see.

The magnetic cycles

The sunspot numbers reflect changes in the magnetic activity of the Sun. When graphs are made of sunspot numbers, slight variations of the Sun's magnetic field (the solar magnetic flux) move in sync with the sunspot numbers. While this pattern is clearly cyclical, out in space and around all the planets this magnetic field is not symmetrical. After all, the Sun is spinning and the polarity of the field is switching every 11–13 years. There are also huge explosions on the Sun's surface (solar flares) that throw charged particles into space and that further complicate the turbulent structure of the Sun's magnetic field.

The Sun's magnetic field was well described in the *NASA News* of June 9, 2009:

> ...the Sun's magnetic field extends all the way to the edge of the solar system. Because the Sun spins, its magnetic field becomes twisted and wrinkled, a bit like a ballerina's skirt. Far, far away from the Sun, where the Voyagers are now, the folds of the skirt bunch up. When a magnetic field gets severely folded like this, interesting things can happen. Lines of magnetic force criss-cross, and "reconnect". The crowded folds of the skirt reorganise themselves, sometimes explosively, into foamy magnetic bubbles. We never expected to find such a foam at the edge of the solar system, but there it is!" says Opher's colleague, University of Maryland physicist, Jim Drake.

> Theories dating back to the 1950s had predicted a very different scenario. The distant magnetic field of the Sun was supposed to curve around in relatively graceful arcs, eventually folding back to rejoin the Sun. The actual bubbles appear to be self-contained and substantially disconnected from the broader solar magnetic field.

The case of cosmic rays is illustrative. Galactic cosmic rays are subatomic particles accelerated to near-light speed by distant black holes and supernova explosions. When these microscopic cannonballs try to enter the solar system, they have to fight through the Sun's magnetic field to reach the inner planets.

Opher points out, "The magnetic bubbles appear to be our first line of defence against cosmic rays. We haven't figured out yet if this is a good thing or not. On one hand, the bubbles would seem to be a very porous shield, allowing many cosmic rays through the gaps. On the other hand, cosmic rays could get trapped inside the bubbles, which would make the froth a very good shield indeed."

This complex, bubbly and sometimes porous magnetic field varies in sync with the 11–13 year sunspot cycles. Over the centuries, it also waxes and wanes in sync with long-term trends in sunspot numbers and also slight changes in sunspot cycle length. Such numbers can increase over time and occasionally there will be years without any apparent sunspots at all. For example, the second half of the 20th century saw what many called a Solar Grand Maximum when the maximum number of sunspots in each cycle increased throughout the second half of the 20th century. In the 17th century there was a period of weak solar activity when the opposite occurred and there were no sunspots for many years.

The solar waistline

Continuous nuclear explosions tend to expand the Sun while gravitational forces cause the Sun to contract. Hence, the exploding Sun is a pulsating throbbing body and, like a throbbing human heart, there are regular changes in the Sun's radius and therefore in its surface area. Russian scientists have been tracking changes in the solar radius using a special instrument on the International Space Station (Diagram 11.1). Descriptions of this experiment, called Astrometria, are available on the Internet.

The Russians have been measuring changes in the Sun's radius since the early 1990s. They report a change in the Sun's radius of 130 km in each sunspot cycle. The shorter solar radius corresponds with high numbers of sunspots at the peak of each sunspot cycle (Qu and Xie, 2013). Given that the Sun's radius is around 692,500 km, this 130 km variation equals a change in the surface area emitting heat to space of around 565 million sq km; an area slightly more than the surface area of the Earth. The Russian research is also suggesting a change in radius around 300 km for each 208 year sunspot cycle (De Vries cycle) that equals a surface area change of 1.3 billion sq km; an area nearly three times the surface area of the Earth. The Russians argue that such changes in surface area slightly vary the heat radiated from the Sun that reaches the Earth.

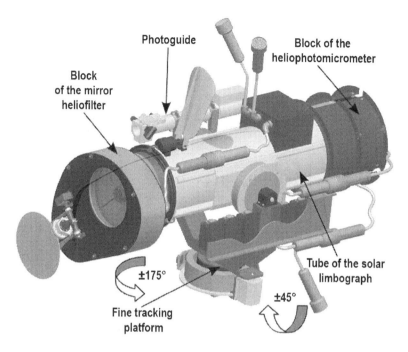

Diagram 11.1: A diagram of the Russian instrument measuring solar diameter attached to the international space station (www.gao.spb.ru/ english/astrometr/index1_eng.html).

This instrument tracks small changes in solar radius, that in turn reflect changes in the Sun's surface area. The surface area of a sphere varies according to its radius squared. Consequently, all things being equal, more heat is emitted from a larger surface area than a smaller surface area. A 100 km change in the Sun's radius is equivalent to an extra 500,000 sq km radiated heat into space; an area equal to the surface area of the Earth.

The Russian research clearly describes a never-ending battle between the explosive and gravitational forces on the Sun. Now, new research from Britain takes this internal disequilibrium a step further. Professor Valentina Zharkova, a Professor of Mathematics at the Northumbria University, and her colleagues have detected two slightly different 11 year magnetic cycles inside the Sun. They proposed one magnetic cycle originating in the outer section of the Sun and another in its innermost core. They detected slight phase differences between these dynamos and reported that these differences can be used to predict sunspot numbers. *Zharkova and her colleagues predict that sunspot numbers will continue to drop and that by 2030–2040 sunspot numbers will be as low as during the Maunder Minimum in the 17th century – a cold period known as the Little Ice Age.* So they predict no sunspots and a cold period around 2030–2040 when these two magnetic cycles will be completely in phase (Zharkova, 2015).

Solar wobbles

While the Sun's gravity controls the orbits of its planets, those planets in turn cause small orbits or wobbles in the Sun's position. The main effect comes from the largest planets, Jupiter and Saturn. Professor Rhodes Fairbridge, a famous Australian scientist, who worked at Columbia University, New York for many years, describes the slight variations in the Sun's heat reaching Earth caused by these small changes in distance between them (Fairbridge and Shirley, 1987).

The Sun wobbles in a three dimensional space up to twice the Sun's diameter. Such wobbles quickly vary the Sun's distance from Earth, that is around 150 million kilometres, by more than one million kilometres. The orbital centre of these movements is called the barycentre and the complex motions around the barycentre repeat every 179 years in what is called an epitrochoid cycle. Diagram 11.2 shows the Sun's positions for the 50 years from 2000 to 2050. As with all cycles, no one cycle will be an exact repetition of another. According to analysis from the Pulkovo Observatory in Russia, the past four epitrochoid cycles all began with very cold periods.

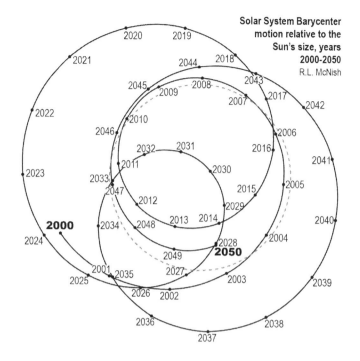

Diagram 11.2: The projected path of the sun through the period 2000 to 2050 due to the gravitational pull of the various planets (Wikipedia).

There has been speculation for years as to a possible link between various cycles of the largest planets, Jupiter and Saturn, with the Sun. This epitrochoid cycle repeats every 179 years and is close to two solar diameters. Scafetta points out that these orbital variations cannot be left out of the climate debate (Scafetta, 2006).

The scientific examination of the barycentre has not been helped by frequent references to Jupiter and Saturn in astrological writings. There was the book written by Dr John Gribbin (an astrophysicist at the University of Sussex, England) and a Stephen Plagemann that was published in 1974 called *The Jupiter Effect*. Their predictions of an increasing number of earthquakes and volcanic eruptions were based on the close co-incidence of the 11.87 year sidereal period of Jupiter, the mean sunspot cycle of 11.3 years and a line-up of planets. This scaremongering sparked panic and caused various doomsday sects in the US to head for the hills in 1982.

UV-rays at work

The UV radiation emanating from the Sun covers a wide spectrum, much like visible light. The spectrum ranges from longer UV rays near the violet end of the visible light spectrum to extremely short UV rays close to X-rays. During the 11–13 year sunspot cycles the UV radiation varies by 6% (longer wave UV) to 75% (shortest wave UV). Those wavelengths close to light are absorbed in the Earth's lower atmosphere by aerosols, clouds and gases such as nitrous oxide. The middle range of UV radiation forms ozone in the stratosphere and the shortest waves of UV are active in the outermost reaches of the atmosphere, forming a band of charged particles around the Earth known as the ionosphere.

The stratosphere, where most ozone is formed, is the zone in the Earth's atmosphere that is immediately above the troposphere, and the troposphere is the zone of the atmosphere closest to the Earth's surface. The troposphere is 20 kilometres deep at the equator and only 7 km deep at the poles. Temperatures drop throughout the troposphere to between -45°C and −75°C, but in the stratosphere temperatures begin to warm again and are around −3°C at heights of around 50 km.

The ozone formed in the stratosphere is a minor greenhouse gas. The Sun's UV rays strike molecules of oxygen that have two oxygen atoms and, on the bust–up, ozone is formed where single atoms of oxygen re-combine to form a molecule with three oxygen atoms. There is a theory that this ozone can cause high altitude cloud formation, and this would significantly vary the amount of solar radiation reflected into space. Yet another way our Sun affects the Earth's climate.

In 1998, Dr Joanna Haigh of the Imperial College, London, proposed that a 0.1% change in UV radiation could cause a 2% change in ozone concentration:

> ... the proposed amplification mechanism works by having solar UV irradiance (wavelengths less than 0.4 microns) heat the lower stratosphere and upper troposphere, and significantly modulate the amount of upper-atmospheric convection, ice cloud cover and ozone concentration, which consequently alter the transparency of the terrestrial atmosphere to solar short wave radiation (Haigh, 1998).

Dr Willie Soon of the Smithsonian Institute, Washington has also written extensively on the role of ozone in the stratosphere. He contends that ozone formed by middle range UV radiation from the Sun could have a role in world climate by influencing the formation of high cirrus clouds and generating a disproportionate climate response (Soon et al., 2000).

Such hypothetical arguments about the upper troposphere and lower stratosphere are developed through modelling and need to be verified by data that is difficult to obtain. These areas are too close to the Earth to position stable, orbiting satellites. Most UV measurements are recorded by instruments attached to radiosonde balloons and even these can only penetrate the lower stratosphere.

There are also various satellites, well above the stratosphere, measuring incoming UV radiation from the Sun. The TIMED satellite was launched in 2001 and for two years explored the atmosphere in the 60 to 180 km range. The SORCE satellite, with an orbit height of 645 km, was launched in 2003 and, apart from measuring incoming solar radiation (TSI), it has a UV monitor (called the XUV- Photometer System). Although SORCE was planned for a five year life, the satellite is still operating. The plan to replace SORCE in 2011 failed and a replacement is long overdue.

So, changes in stratospheric ozone, changes in the Sun's magnetic field, changes in sun spots and small changes in distance between the Earth and the Sun show there are multiple connections between the Earth and the Sun that affect the radiation reaching the Earth.

Total Solar Irradiance

The heat given out by the Sun is in the form of short-wave radiation and is called its Total Solar Irradiance (TSI). This incoming radiation is measured by instruments called cavity radiometers in watts per sq metre. These instruments are fitted to satellites orbiting well outside the atmosphere of the Earth. The readings are taken by sensors at right angles to the incoming radiation. The heat radiation given out by the Sun, even during short 11–13 year sunspot cycles, as read by these instruments, does not have wild short-term fluctuations like those of the magnetic field, nor is this radiant energy like UV rays and X-rays that show a wide range of values during those cycles.

The cavity radiometers measuring TSI degrade due to exposure to light and due to the bombardment of cosmic rays. The readings also have small variations due to gradual decline in a satellite's orbit. Even the satellite SORCE launched in 2003 had to go on restricted battery mode in October 2012 to protect its sensors and has been nursed along since then.

There are disagreements about the exact amount of solar radiation reaching the Earth's outer atmosphere. Estimates of TSI from various satellites vary from 1360 to 1365 watts per sq m. Even the Picard spacecraft measured TSI in 2010 summer to be 1362.1 watts per sq m with an error uncertainty of ±2.4 watts per sq m. This estimate covers a range from 1359.7 to 1364.5 watts per sq m (Meftah and others, 2014).

Because the differences in TSI measurements are small, it is difficult to determine what changes are due to instrument variations and what changes are actually due to the Sun. One school of scientists holds that changes in TSI are so small that recent global warming must be due to processes on the Earth such as changing greenhouse levels. Other scientists hold that small changes in TSI are significant and have to be linked with all the other connections between the Sun and the Earth.

Interpretations of TSI have been compounded by a gap in the satellite data. In 1986 NASA failed to place a new satellite into orbit as the shuttle taking it into orbit exploded (the tragic Challenger disaster when seven astronauts lost their lives). Without this new satellite there was a gap in the satellite measurements that led to very different interpretations of trends in TSI for the next decade.

One group of scientists led by Dr Mike Lockwood of the University of Reading, England corrected the satellite TSI data using their interpretation of instrument degradation. The data (often referred to as the PMOD data) come from a Swiss-funded research institute used by the IPCC called the Physikalisch-Meteorologisches Observatorium Davos. Their *corrected* data show a slight downturn in TSI and indicated that the Earth should have cooled, not warmed in that period. For many scientists this meant that any surface warming in the 1990s must have been caused by an increased greenhouse effect in the atmosphere and not by the Sun. For them, this was proof of global warming due to increasing greenhouse gases (Lockwood and Frolich 2008, Gray et al., 2010).

Other physicists, such as Professor Nicola Scafetta and Dr Bruce West from Duke University, South Carolina disagree with the instrument degradation assumptions of PMOD. Their interpretations of TSI are opposite to that by PMOD and show a slight increase in TSI in the same period (Scafetta and West, 2006). There is also another set of data collected by the Royal Meteorological Institute of Belgium that is in agreement with Scafetta and West and at odds with the PMOD data.

It should also be noted that Russian scientists from the Pulkovo Observatory interpret some of the warming in the lower atmosphere during the final decade of the 20th century a little differently. For them there is some warming due to a natural lag in the Earth-Sun energy relationship during which the oceans give out heat to maintain thermal balance in the Earth's climate system. They state:

> Here, despite the fact that the amplitude of TSI variation is approximately 0.07% during the "short" 11-year cycle, its influence on climate is softened by the thermal inertia of the ocean. But if an increase or decrease of the TSI variations amplitude lasts for two subsequent cycles given a similar course of its two-century component, the climate will eventually change correspondingly, but with a delay of 15 ± 6 years caused by the thermal inertia of the ocean (Astrometria internet site, 2015).

In 2000, Tobias and Weiss criticised assumptions by the IPCC that small changes in the Sun could not influence climate:

> The IPCC dismissed any significant link between solar variability (Total Solar Irradiance) and climate, on the grounds that changes in irradiance were too small. Such an attitude (of the IPCC) can no longer be assumed.

And *NASA Science News* in 2013, even though supporting the narrow view of TSI held by the IPCC, reports that the situation is not so simple:

> In the galactic scheme of things, the Sun is a remarkably constant star. While some stars exhibit dramatic pulsations, wildly yo-yoing in size and brightness, and sometimes even exploding, the luminosity of our own Sun varies a measly 0.1% over the course of the 11-year solar cycle. There is, however, a dawning realization among researchers that even these apparently tiny variations can have a significant effect on terrestrial climate. A new report issued by the National Research Council, 'The Effects of Solar Variability on Earth's Climate', lays out some of the surprisingly complex ways that solar activity can make itself felt on our planet (NASA Science News Jan 8, 2013).

The margins of error in the instruments measuring TSI have already been mentioned. For instruments on the Picard satellite they were ±2.4 watts per sq. metre. This means that estimates of global warming over the past 150 years in the IPCC reports are inside the error–margins of the instruments measuring TSI. So what meaning can be attributed to statements by the IPCC that the extra warming caused by increasing greenhouse gases over the past 150 years has been 2.3 watts per sq metre? Even the instrument error bars are around that level, and the range of estimates of TSI from various satellites varies by at least five watts per sq m (Soon, in Moran, 2014).

Who is the boss?

While the variations in TSI are small in percentage terms, they could be extremely significant. This means that the critical key to the climate debate is the equilibrium climate sensitivity index (ECS). In the modelling chapter (Chapter 10), the climate sensitivity index was explained. The ECS is an estimate of the temperature rise in the Earth's lower atmosphere if carbon dioxide levels double from their value in the mid-19th century. The average ECS used by the IPCC is 3.0°C; a value estimated by the American Academy of Science at a meeting in 1979.

At the present time the Earth's average atmospheric temperature is about 15.0°C. This temperature is also expressed in degrees Kelvin when temperatures are measured from absolute zero. A degree Kelvin is the same as a degree Centigrade, so 0°K is -273°C and 0°C is 273°K. This means that 15.0°C and can also be expressed as 288°K.

An ECS of 3.0°C leading to a doubling of carbon dioxide levels increases the Earth's temperature from 288°K to 291°K, an increase of 1%. But, estimates by other scientists of a lower ECS around 0.5°K (0.5°C) indicate a temperature rise of only 0.17% if carbon dioxide levels double.

Within the computer climate models, the high ECS used by the IPCC puts rising greenhouse gases in the driver's seat, so that:

- It is virtually impossible to have temperature pauses as greenhouse levels rise.
- Computer model programmes will wrongly predict an accelerating temperature trend as greenhouse levels rise.
- Temperature rises caused by increasing greenhouse gases will be so dominant that small changes in the Sun's orbit, or changes in the activity of ozone in the stratosphere, or variations in the Sun's magnetic field, or small variations in TSI, will have little effect on world temperature trends.

But, a lower ECS changes the whole climate ball game, and puts the Sun back in the driver's seat. The climate models are now more sensitive to small changes. Now:

- Small variations in radiant heat energy will affect world temperatures.
- Ozone, formed by middle range UV radiation from the Sun, could have a role in world climate by influencing the formation of high cirrus clouds and generating a disproportionate climate response.

- Small changes in distance between Earth and Sun, due to wobbles of the Sun around the barycentre in the 179-year epitrochoid cycles, could cause slight changes in TSI.
- Small changes in the Sun's surface due to changes in radius could cause slight changes in TSI reaching the Earth's outer atmosphere.
- Cyclical changes in the Sun's magnetic field could either increase or decrease cosmic radiation reaching Earth and therefore increase or decrease cloud cover and either cool or warm the Earth's lower atmosphere.

History speaks

What does history tell us? The satellite measurements of TSI cover only the past 30–40 years; insufficient time to properly understand the effects of small changes in TSI and other possible changes in the Sun's relationship to the Earth's climate. Indeed, the history of sunspots and also the analysis of isotopes formed by cosmic rays in ice cores, tree rings and sediments provide strong evidence of the Sun's influence and the Sun's dominant role in the Earth's climate.

The sunspot evidence

There is a strong correspondence between very cold periods, low sunspot numbers, high cosmic ray activity and weak solar magnetic fields. Diagram 11.3 clearly shows an absence of sunspots in the cold period in the 17th century called the Maunder Minimum. This period is often referred to as the Little Ice Age (a poor term as the name implies the formation of mid-latitude ice sheets). It certainly was a colder period. The winter ice festivals on the Thames River in London during this period are well recorded. The most famous fairs were during the Great Frost of 1683–1684 when the Thames froze over for two months. In China, the great drought and famine in Northern China from 1628–1648 may have contributed to the fall of the Ming Dynasty.

The same diagram shows very low sunspot numbers in another short cold period at the start of the 19th century called the Dalton Minimum between 1795 and 1825. During this period Napoleon invaded Russia, and the extremely severe winter in 1812 was a large factor in the decimation of, not only his 400,000 troops, but also 50,000 horses.

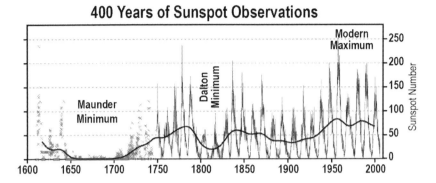

400 Years of Sunspot Observations

Diagram 11.3: Sunspot numbers over 400 years (from Wikipedia).

The relationship between low sunspot numbers and the cold periods of the Maunder Minimum (1645–1710) and the Dalton Minimum (1795–1825) stands out. Many scientists are now predicting low sunspot numbers in the mid 21st century with a possible cold period similar to the Dalton or Maunder Minimum.

The Maunder and Dalton Minima and the detailed history of sunspot cycles over the past 400 years have given climate scientists a key to look for evidence of similar cold periods in the past. During the Maunder and Dalton Minima, when sunspots were low or absent, sediments and tree rings around the world, and ice cores from polar regions, had higher levels of an isotope of carbon, C14 and an isotope of beryllium, Be10. Such isotopes cannot be formed by solar radiation, but only by cosmic radiation travelling near the speed of light. When such radiation (a pot-pourri of atoms, protons, neutrons, and weird sub-atomic particles) hits nitrogen and oxygen atoms in the atmosphere rare atoms of beryllium and carbon are formed. These have different numbers of neutrons from the common forms of carbon and beryllium. This cosmic radiation comes from exploding stars and galaxies. Since the Sun's magnetic field stops much of this cosmic radiation from reaching the Earth, a larger cosmic ray count is evidence of a weaker solar magnetic field. Here was a key to find similar cold periods in the past.

There have been famous pioneers in the field of isotope chemistry who have added to our knowledge of past climates. Sir Nicholas Shackleton at Cambridge pioneered the use of oxygen isotopes in fossils to measure past temperatures. Professor Minze Stuiver founded the Quaternary Research Centre at the University of Washington in Seattle and, over a long career, developed the analysis of carbon 14 in tree rings from

1961 onwards. Professor Willi Dansgaard (Copenhagen, Denmark) and Professor Hans Oeschger (Bern, Switzerland), using techniques developed by Shackleton, completed the first analysis of oxygen isotopes in Greenland ice cores from 1983 onwards. And in the 1990s Dr Gerard Bond from the Lamont-Doherty Observatory, Columbia University, New York pioneered the analyses of beryllium 10 and carbon 14 isotopes in Atlantic sediments.

The cold periods of high cosmic ray bombardment, when more Beryllium 10 and carbon 14 are formed, are normally about 70–90 years long. Apart from the Maunder Minimum 1645 AD to 1710 AD, other earlier cold periods were the Wolf Minimum 1280 AD to 1350 AD and the Sporer Minimum 1460 AD to 1550 AD. There have been shorter 30–40 year cold snaps. The Oort Minimum occurred within the Medieval Warm Period between 1040 and 1080 and the Dalton Minimum between 1795 and 1820 (Miyahara et al., 2009).

The concurrence of low sunspot numbers in the Maunder and Dalton Minima suggests that *low sunspot numbers could be a warning of an impending cold period*, and this has led to much speculation about the future climate of the 21st century. Between 1986 and 2017 there has been a continual drop in sunspot numbers. During that period sunspot numbers at the crest of the three sunspot cycles have been between 150 and 200, then 100 and 125, and then between 60 and 100; clearly a declining trend (Figure 11.4).

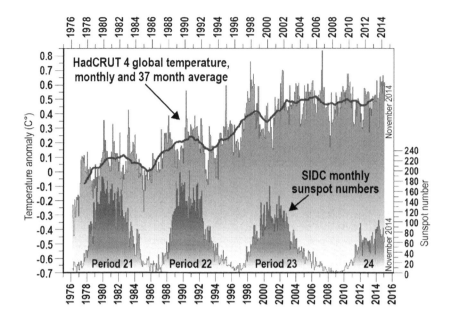

Diagram 11.4: The declining fluctuations of sunspots numbers in the four sunspot cycles since 1976 and the global temperatures known as HadCRUT4 measured by the British Hadley Centre (from Professor Humlum-Climate 4You website).

Note the declining sunspot numbers between the crests of cycles 21 to 24 and the levelling of world temperatures from 2000–2015. The dramatic decline in sunspot numbers is apparent. Further decline in sunspot numbers in future cycles 25 and 26 would be in line in with scientific predictions of a cooling event in the middle of the 21st century. Russian scientists predict a lag between the decline in sunspot numbers and a decline in world temperatures due to an initial heat outflow adjustment from the oceans. It should be noted that the upper 10 metres of the ocean have the equivalent heat content of the total atmosphere; so such a proposed lag is totally reasonable.

The magnetic cloud switch

Scientists are now looking for the mechanism that could explain the clear concurrence of high cosmic ray flux, high carbon 14 and high beryllium 10 isotopes, lower solar magnetic flux and cold climates.

There is a possible link between cosmic rays and cloud formation that could explain these cold periods. If cosmic rays could increase low cloud formation and these clouds reflect more radiation into space, then cosmic rays could cool the Earth. This would put the Sun back in charge because the magnetic field of the Sun would then be the switch controlling the number and intensity of cosmic rays hitting Earth.

A world expert on cosmic rays is Professor Henrik Svensmark, a physicist at the Danish Space Institute in Copenhagen. For Danish scientists, such as Svensmark, variations in this complex, whirling magnetic field vary the number of cosmic rays reaching Mother Earth. According to Svensmark, cloud droplets can form around microscopic atmospheric particles electrically charged by cosmic ray collisions. The nuclei to start a cloud droplet are around 1/5000 of one millimetre and are called Cloud Condensation Nuclei (CCNs). Such nuclei can be minute dust or salt particles and even large organic molecules from gases emitted from plants. Svensmark argues that cosmic rays could increase the number of charged particles that act as CCNs and so increase the amount of cloud, especially low cloud.

Such a mechanism places the magnetic field of the Sun back in charge of the Earth's climate, not so much through changes in the Sun's heat output (Total Solar Irradiance), but through the Sun's magnetic field acting as a switch slightly modulating and amplifying cloud formation and the amount of solar radiation reflected into space (Svensmark et al., 1997, Svensmark, 2007). Recent comments on the amplification factors in the climate system, due to the actions of cosmic rays in cloud formation, are in an article by Yu and Lao (2014).

Many scientists are following this Danish research with interest because of critical problems with the proposition that carbon dioxide has been the main mechanism causing global warming during the past 150 years.

There has been much debate about cosmic rays and climate, and various scientists are of the view that Svensmark and his Danish researchers have not proved their cosmic cloud hypothesis. The trouble is that CCNs are formed near the atomic and molecular level. Other chemicals in the atmosphere could be involved such as sulphuric acid, organic vapours from vegetation, and even organic compounds formed by plankton in the ocean and released into the ocean by sea-spray (Wilson et al., 2015). While some CCNs have been formed in cloud chamber experiments, the exact mechanism may be very difficult to unravel due to the difficulty of creating realistic experimental conditions at this *'near-atomic'* level.

We could easily be in a situation with respect to cosmic rays and cloud formation, as geology was in the early 20th century, trying to explain the configuration of the continents. In 1912 Alfred Wegener presented empirical evidence that continents were once joined. This was further elucidated in his 1915 book, *The Origin of Continents and Oceans (Die Entstehung der Kontinente und Ozeane)* and in various publications until his unfortunate death on the Greenland ice sheet in 1930. However, his theory was not accepted for 50 years until the volcanic mid-Atlantic ridge was mapped using data collected by Atlantic convoys during the Second World War. Then the existence of an oceanic spreading centre was proved and his continental drift theory was accepted as valid, even though the deep mechanism behind such crustal movement is still not fully understood.

The solar solution

The seemingly *idle* Sun has been a poker-faced illusionist that distracted scientists to look elsewhere and they wrongly focused on changing carbon dioxide levels as the key to recent climate change. This *misdirection* meant that the climate debate was side-tracked by focusing on a simplistic, carbon dioxide theory full of inconsistencies.

It may take some future shocks, such as global cooling, to bring the scientific world back to its senses and to admit that in the past 40 years an exaggerated role has been ascribed to changing carbon dioxide levels and that it has not appreciated the complexity of the climate system and the role of our Sun.

Looking at the Sun that so dominates the solar system we may produce multiple explanations that describe synchronicities in events that seem to fit the data, without being able to prove which, if any, are valid. After all, interrelationships due to thermal, gravitational or magnetic forces between the Sun, the galaxies of the cosmos, the Earth and the Sun's other planets have had over five billion years to develop. Such relationships may have been amplified by resonance as synchronous periodicities were established. So, due to the nonlinear nature of such complex systems, the exact causes of such periodicities may remain unknown, and any complicated mathematical explanations for such periodicities may indeed be mathematical accidents, not solutions.

It is clear from climate history and from a careful analysis of the relationship between carbon dioxide levels and temperature over the past 150 years that the Equilibrium Climate Sensitivity index is much lower than estimated by the IPCC. A low estimate means that other factors could cause small changes in the solar irradiance reaching the Earth's surface and help explain short-term changes in climate.

Indeed, when the Equilibrium Climate Sensitivity index is low, a radical argument can be made that an important key to global climate history may be found in those sunspot-free cold events that were caused by significant decreases in the Sun's magnetic field, such as the Little Ice Age in the 17th century. It may take hundreds of years for the Earth to recover from such cold trauma. This recovery period would be a return to homeostasis, much like the human body recuperating from severe illness. In this interpretation such convalescence from severe trauma does not require special significant warming events. During this recovery, small changes in solar heat output, or further small variations in solar radii, or small variations in ozone levels, or further small variations in cosmic radiation, or small heat exchanges between the oceans and the atmosphere may each easily cause pause-warm patterns to overlay a natural upward, warming, recovery cycle.

We have accepted the magician's dare and found that the Sun has many tricky ways to influence the Earth's climate.

As a master illusionist, the Sun probably has some more tricks up its sleeve, but at the very least we acknowledge the Sun is king and a major factor in recent climate change.

Chapter 12.

Other climate shocks

In the looking glass

There has been some warming effect from increased greenhouse gases, but not enough to override various temperature pauses in the past 150 years. It is clear that a warming cycle began over 300 years ago in the early 18th century. Such a warming cycle, in its timing and general shape, is in rhythm with the Medieval, Roman and Minoan Warm Periods. In pop-music parlance, the recent climate changes don't make the charts. Humankind is presently living in a benign climate epoch. As James Lovelock remarks:

> What we are doing to the Earth by our industry has already changed it, and more change is likely to happen, but nothing so far justifies the frantic cries of the environmental activists, who amplify such hyperbolic cries as: 'And it will destroy all life on Earth!'. So let us keep our cool as the Earth gently warms, and even enjoy it when we can (Lovelock, 2014).

Historical evidence shows that past warming periods have been periods of prosperity with increased agricultural production and population growth, and that this warm period has been the warm period *par excellence* for human development. Compared to dramatic climate events in the past, recent changes in climate are not *unprecedented*.

There has been so much panic. One of the worst examples was a United Nations Environmental Program report predicting that, due to increased global warming and rising sea levels, there would be 50 million climate refugees by the year 2010. This was climate alarmism at its worst. The decade ending in 2010 saw record agricultural production, a pause in world temperatures, and no acceleration in sea level rise. The map depicting those migrations was quietly withdrawn.

The alarmism about rising greenhouse gases is totally misplaced. But could there be other climate shocks? Possible future catastrophes, while not the focus of this book, are more related to population stress, economic failure, and biological disasters from rogue strains of viruses or bacteria, or to human self-destruction through the use of biological or nuclear weapons. The mass migrations taking place at the present time are economic or political in nature from poor, impoverished, unstable, unsafe or overpopulated regions towards wealthier, safer, politically stable zones.

Other shocks

But, are we safe? It is interesting to put the current world obsession with rising greenhouse gases in perspective and examine dramatic events that could lead to climate disasters such as:

- global cooling.
- meteor or comet strikes.
- gigantic solar flares.
- volcanic eruptions.
- catastrophic melting of large frozen methane deposits.
- super-floods.

Global cooling

There was a rapid cooling of the Earth 12,800 years ago. It was the beginning of a period now called the Younger Dryas. There is disputed evidence that this cooling was the result of a comet impact with Earth. In any event the change was rapid, within a decade, and something quite outside any recent climatic event.

There is clear evidence that sun-spot numbers are declining. Low sun spot numbers (as already explained in the Chapter 11) have been associated with cold events in the past. What population migrations could occur if global cooling affected the Northern Hemisphere wheat belts and the agricultural production of northern China? In the past, it was increasing cold that led to historical migrations from the Eurasian steppes into Europe after 300AD or from Siberia across Bering Strait to North America during the most recent ice age.

Looking at the sunspot data new prophets are emerging with forecasts that adjust their doomsday narrative to a cooling Earth. Now, their prophecies are not of rising sea levels and a baked planet, but of freezing conditions, famine, disease and drought.

One such prophet is David Archibald. His book, *The Twilight of Abundance*, has excellent analyses of sunspot data, but Archibald's writings link the climate catastrophe of global cooling, as indicated by his sunspot analysis, with dire predictions of oil and gas shortages, possible world wars, possible serious famines and certain religious texts (Archibald, 2014). Archibald sees sunspots as being able to provide a climate forecast for hundreds of years, whereas nonlinear science indicates that any one climate changing force can be over-ridden by others and therefore cannot be used as a one-stop forecasting tool. Furthermore, the absolute unpredictable quirkiness of the system is always present. The whole basis of science is its own internal logic. It would seem that both sides of the climate debate are breeding their own charismatic prophets.

Apart from sunspots, dramatic changes in ocean currents could cause global cooling in some regions. An overturning of deeper, but warmer heavier salty water in the Arctic is a ticking time bomb, but it is not clear what could cause that water to upwell and melt all the Arctic ice. Or again, the warm Gulf Stream flows from Florida and the warm West Indies diagonally northwards towards Europe. If that ocean current failed, then northern Europe would experience a frigid climate like that of Hudson Bay in Canada at the same latitude. It is possible that freshwater from melting ice sheets shut this current off 10,000 years ago, but there is no evidence of any impending collapse at the present time.

In the past some cooling did have dramatic consequences. For example, crop failure often led to famine because grain could only be transported over short distances. Today modern transport can alleviate that type of stress, and modern technology has made it far easier to live in cold climates and adapt grain to thrive in harsher conditions. However, cooling would still shift Northern Hemisphere agricultural belts southwards, while a larger polar high pressure system in the Southern Hemisphere would push low pressure systems northwards so that the southern regions of Australia, Argentina and South Africa should all enjoy higher rainfall that could transform many semi-arid areas into fertile plains.

The next ice age will cause disastrous cooling, yet its timing is difficult to predict. One cannot just use the latest short interglacial period of around 17,000 years as a guide because the length of each interglacial depends on the interplay of the three variations in the Earth's orbit, outlined by Milankovich and Croll over 100 years ago. The lengths of recent interglacials have varied from short, around 13,000 years, to long, around 28,000 years.

A group of seven scientists from England, America and Germany led by Professor Chronis Tzedakis from University College, London think the present interglacial will be of the shorter variety. Something scary, but not in our time (Tzedakis et al., 2012). So sometime between 1000 and 16,000 years from now, humankind will have to face an ice age. Then, northern Europe to the Spanish border, most of Russia, Siberia and Manchuria, and North America to just south of New York, will be covered by 1–2 kilometres of ice. That will be a doomsday scenario hard for us to contemplate as billions of people will be displaced. Iconic features reflecting the height of human endeavour, such as the Great Wall of China, the stunning medieval cathedrals in Germany and France, the Hermitage Museum, the great skyscraper complex of New York, will one day collapse under the weight of ice. The Sahara desert will once again be a huge fertile grassland with large rivers and abundant lakes. The centre of the habitable world will shift to the Tropics and Southern Hemisphere, as south of the equator the only ice sheets will be in Southern Chile (Patagonia) and Antarctica.

Meteor and comet strikes

Meteor and comet strikes have been an integral part of the history of our solar system. The pock-marked Moon bears witness to their destructive power. They are still around, but not so many. Over billions of years, the gravitational pull of the large planets, such as Jupiter and Saturn, has acted like a vacuum cleaner and pulled many into the outer reaches of the solar system to the Oort Belt. Meteors can strike at enormous speeds (say 60,000 km/hr) and even a small meteor 100 metres across has the destructive power of hundreds of atomic bombs. As a meteor plunges into Earth, vast quantities of dust and water are thrown into the atmosphere. Any dust that reaches the upper troposphere and lower stratosphere can spread throughout the upper atmosphere and stay there for years. There it will reflect solar radiation and cause global cooling for many years.

If a meteor or comet is on a collision course with Earth, there is not much we can do, unless a satellite intercepts and destroys it in outer space.

The force of a meteor strike was dramatically shown when one, thought to be only 60–100 metres in diameter, blazed across the Siberian sky in 1908. It was not a direct strike and it vaporised in the atmosphere. The shock waves flattened 80 million trees in an area of 2000 sq km. This event is known as the Tunguska Event (Diagram 12.1). If such an event occurred near a modern city no building would be left standing.

Diagram 12.1: There are many old black and white photos of the devastation left by the meteor that vaporised in the Siberian sky in 1908. The incident has been named the Tunguska Event.

Solar flares

Solar flares are due to large explosions in the outer layer of the Sun. When these occur large plumes of flame flare outwards from the Sun's surface for millions of kilometres. These solar flares bombard the Earth with billions of electrically charged particles.

The largest known solar flare occurred in 1859 and led to what is called the Carrington Event. The solar flare was of such intensity it lit up the night sky, powered telegraph lines on its own accord and would have caused most of the world's present electrical transformers to burst into flames. The cost of a Carrington event today would be trillions of dollars, economic chaos and probably worldwide recession; but there is no evidence that such an event would trigger sudden or significant climate change, even though it would be a catastrophic event.

Volcanoes

Volcanoes have been part and parcel of human history and are etched in the human imagination. The Mediterranean region has seen some huge eruptions that have destroyed ancient civilisations, such as Santorini

3600 years ago. Eruptions such as Krakatoa in Indonesia (1883) and Mt Pinatubo in the Philippines (1991) created dusts clouds that did affect climate in bands around the Earth for a few years. The largest recent eruption was Tambora in Indonesia in April, 1815 during the cold Dalton Minimum that began in 1795. Although it did not initiate this cold period, Tambora caused the famous *year without a summer* in northern Europe in 1816, when snow fell in England during July. While these events caused some climate change at their latitude for a few years, they had no long-term effect.

With respect to volcanoes, there is one type of event that could affect world climate. Underneath the Earth's outer crust, which is 30–50 km thick, there is an area called the upper mantle upon which the crust slides. At various times in the Earth's history huge plumes of lava can be ejected from this upper mantle and break through hundreds of fissures on the Earth's surface. The most famous event of this type occurred in India on the Deccan Plateau about 60 million years ago. Lava was ejected over an area of over 500,000 sq km and, in places, the lava flows were two kilometres deep. The scale of this type of event, ejecting large volumes of dust and sulphur dioxide over thousands of years, would cause cooling on a global scale as the volume of gases and dust is so large that a significant amount would form a blanket around the Earth in the upper troposphere and lower stratosphere. Such a blanket could cool the Earth for thousands of years.

One region that could face such an event today is the Yellowstone National Park in the United States, where there is evidence for an underlying hot mantle plume. An outpouring of lava from such a plume would be accompanied by gas and dust clouds that would rapidly cool most of the United States, completely destroying its agricultural potential, and rendering much of its area unliveable.

Nevertheless, there is no evidence to show when and if such eruptions may take place. Imagine what would happen if such a plume erupted under the large East Antarctic Ice Sheet. On that scale it would not only affect the atmosphere, but also world sea levels.

The frozen methane bomb

Much talk is made of frozen methane. This frozen methane is actually methane hydrate and occurs when methane and water combine at low temperatures to form an ice-like substance. It is formed when natural gas escapes through the Earth's crust and encounters, either water near

0°C in the deep ocean, or moisture in rock and soil at near freezing temperatures. There are trillions of tonnes of this stuff in the ocean floor and in the Arctic permafrost. It is thought this frozen methane exceeds all the known deposits of natural gas. In the Arctic, frozen soil and rock near the surface is called *permafrost*. That permafrost can extend to depths around 1500 metres due to the influence of freezing surface temperatures over thousands of years. During recent Arctic summers, the permafrost thaws only to depths around 30 cm, and then this top layer refreezes again in winter.

Doomsday scenarios abound about sudden, catastrophic methane releases that could effectively poison the atmosphere and kill all animal life. A novel called *Mother of Storms*, incorporating the release of such methane with other science fiction events, was written by John Barnes in 1994.

In fact, methane hydrates are less frightening. A good summary of the relationship between methane hydrates and climate change has been written by Dr Carolyn Ruppel, Chief of the United States Geological Survey Gas Hydrates Project (Ruppel, 2011). This report was prepared on the basis that the IPCC is correct in assuming that increasing greenhouse gases will cause a 0.2°C per decade warming of the Earth (an assumption many scientists are now saying is too high).

Dr Ruppel's main conclusions about methane hydrate are:

- 99% is in ocean sediments and 1% in the Arctic permafrost.
- When methane leaks from the ocean floor it reacts with oxygen in the water to form carbon dioxide, and this carbon dioxide, rather than methane, would bubble into the atmosphere. The destruction of oxygen in the ocean from such a release would be harmful to marine life.
- It is difficult to warm the deep ocean floor temperature to initiate such a release of methane hydrate, as even in the tropics the deep ocean is not far from freezing temperatures. It could take 100 years to have a enough warming effect on the deepest ocean floor to trigger such a release and, even then, that warming would only thaw the upper few metres and it is even difficult to see how warming could reach the waters of the deep ocean.
- The methane hydrate deposits most susceptible to release are those in sediments on the ocean floor in shallow Arctic waters, up to a water depth of 400 metres. There is already evidence of methane release in the Barents Sea near Alaska, the Beaufort Sea north

of Alaska, and the Greenland Sea around Svalbaad. However, the reaction of this methane with oxygen in the ocean prevents most of this methane from reaching the atmosphere. And there is no data showing any significant increase of atmospheric methane in the past 10 years.

- The Arctic permafrost has developed over thousands of years. Global warming of the Arctic can only slowly permeate that permafrost. Due to the low temperatures at those latitudes, it would take up to 1000 years for the warming of rock and soil to extend down to 200 metres, let alone deeper levels.

- There have been some surface methane releases leaving shallow craters in the surface soils of Siberia and North America. While looking dramatic, these releases are not significant in terms of the global climate.

There has been considerable media panic about methane release, but such panic is based on poor information:

Catastrophic widespread dissociation of methane gas hydrates will not be triggered by continued climate warming at contemporary rates (0.2°C per decade, IPCC 2007) over the timescales of a few hundred years. Most of the world's hydrates occur at low saturations and in sediments at such great depths below the sea floor or onshore permafrost they will be barely affected by warming even over 1000 year (Dr Ruppel, 2011).

Super-floods

The most rapid rises in sea level in the past 12,000 years were caused by collapsed ice-sheet lakes. They are well described in the book *The Frozen Earth* (Macdougall, 2006, pp. 89–112). These lakes covered thousands of square kilometres and sat on the top of ice sheets; ice sheets that were 2 to 3 kilometres thick.

The largest of all these ancient, ice-sheet lakes has been called Lake Agassiz and it covered the Great Lakes area of the US and much of central Canada. This ice-sheet lake stretched from Hudson Bay in northern Canada southwards to Minnesota in northern US. Its water content was twice that of the present Caspian Sea. As the world warmed, waters from this lake cascaded over the collapsing ice sheet edges, and the ensuing walls of water would have been hundreds of metres high. Sudden pouring of this fresh water into the Atlantic through the Hudson Bay area could have been enough to disrupt the warm Atlantic Gulf Stream and bring much colder climate to northern Europe. At other times the southern edges of the ice sheet collapsed. Modern day tsunamis caused by earthquakes pale into insignificance compared to these walls of water hundreds of metres high and travelling at speeds well in excess of a hundred kilometres an hour across the landscape.

During the most recent ice age, in some places the mouths of deep valleys filled with water were blocked by ice. In Montana an ice wall blocking such a valley collapsed and the rushing waters created the area called the scablands of Montana (Diagram 12.2). The power, speed and turbulence of walls of water, hundreds of metres high, drilled holes in granite and left wave forms in solid rock. Valleys that normally take millions of years to form were simply gouged out in hours. In Montana, it is calculated that 30 cubic kilometres of water per hour were discharged into the Columbia River area, and then surged towards the Pacific Ocean. The ridges of debris from this event can still be mapped by sonar as they extend for more than 1000 kilometres along the Pacific Ocean floor west of Seattle.

There is evidence of similar super-floods in Sweden, the Black Sea area, central Canada, and Siberia. Indeed, the English Channel itself, which separates France from England, was formed by a super-flood as cascading water from an ice dam in the North Sea made short-shrift of the soft white chalk deposits between England and France, leaving us with the spectacular white cliffs of Dover.

These past events must have left an indelible impression on the humans who saw them and survived. To a person perched on a nearby mountain top what would have been the shaking of the earth and the noise in Montana, as 30 cubic km of water an hour gouged out the landscape? Imagine being near the cliffs of Dover and seeing the land between England and France disappearing before your very eyes. It is not surprising that stories of these super-floods were passed down through generations and were incorporated into the religious cosmology and writings of the Indian, Babylonian and Jewish peoples.

It is difficult to generate floods of such scale today. There are no mid-latitude ice-sheets, and the climate in Antarctica and Greenland is too cold to produce such enormous lakes, unless there were large volcanic eruptions through those ice sheets.

Diagram 12.2: The Montana Scabland (photo-Tom Foster: www.HUGEfloods.com).

The scablands left by the collapse of an ice dam in Montana are a dramatic example of landscapes carved by catastrophic floods related to ice sheet collapse. It is calculated that about 2000 cubic kilometres of water flowed over the landscape during this event in Montana. There are potholes carved in granite and mesas left between the channels. During such an event, the worst of the devastation would have been over in a few days! The source of the water has been named Glacial Lake Missoula.

A final perspective

When we allow our imagination to run riot and examine possible climate catastrophes, the present temperature rise of around 0.8°C over the past 150 years looks pretty trifling. We need to place the present climate debate in perspective. It is a time for realism, not alarmism. Moreover, we should be careful not to forecast a greenhouse catastrophe using the present faulty climate models.

Our increased knowledge of the past should put us on *station alert* looking for future events that could lead to global catastrophes. Such events will probably come from left field. History tells us that eventually there will be some type of a climate catastrophe, but that will be outside the parameters of the catastrophes predicted by the present faulty *greenhouse-driven climate models* used by the IPCC.

Chapter 13.

A final check on the IPCC

The global watchman

The politico-scientific body called the Intergovernmental Panel on Climate Change (IPCC) was formed in 1988 as a merger between the World Meteorological Organisation and the United Nations Environmental Programme. At the time, there was increasing concern about global warming. At its very beginning the IPCC presumed that man was influencing climate and temperature through the emissions of greenhouse gases.

While its charter is to supply global reports on climate, and coordinate a response to 'man-made' climate change, the IPCC does not conduct any climate monitoring in its own right and it relies on advice from scientists appointed to its various committees. The main IPCC reports are called assessment reports and are released every 4 to 7 years. Each assessment reports contain volumes of material and a summary statement. The summary statements are the most widely read; much like government *State of the Nation* reports (IPCC Reports: 1990, 1992, 1995, 2001, 2007, 2014).

From time to time, there are reports on specific problems. For example, a special report on emission scenarios was released in 2000; a special report on renewable energy sources and on climate mitigation scenarios was released in 2011; and also in the same year, another special report was released on managing the risks of extreme events and disasters to advance climate change adaptation (IPCC Reports: 2000, 2011).

The close link between human activity, greenhouse gases and temperature seemed obvious when the first three IPCC Assessment Reports were written. This was the period between 1975 and 1998 that included the strongest warming decades of the 20th century. However, by 2006 it was becoming increasingly evident a *temperature-pause event was in progress*. Even though that pause began around 1998, a few more years of data were necessary to confirm a clear trend.

IPCC presumptions

Right at the start IPCC reports suggested that greenhouse gas emissions would need to be stabilised to halt a rise in world temperatures due to human (anthropogenic) interference with the climate system. A number of future greenhouse scenarios were formulated in the IPCC Fifth Assessment Report in 2014. These scenarios are called Representative Concentration Pathways 2.6, 4.5, 6 and 8.5 (RCPs). Each scenario makes assumptions about the future population of the world, changes in agriculture, changes in methane levels, the timing of possible action to reduce carbon emissions and a trend in future carbon dioxide levels.

The RCP scenario 2.6 to the year 2100 postulates a fall in oil use, a rise in world population towards nine billion, a reduction in methane emissions, a levelling of carbon dioxide concentrations to the year 2020 and declining carbon dioxide concentrations to the year 2100. The RCP scenario 8.5 to the year 2100 postulates an increase in croplands, a population of 12 billion, a heavy reliance on fossil fuels, no carbon reduction, and a 300% increase in carbon dioxide levels.

Many scientific organisations have designed computer models to investigate these various scenarios and their results are used by the IPCC. The IPCC proclaims that in its current reports there are 300 baseline scenarios and 900 mitigation scenarios; these scenarios are critical in all IPCC documents.

The watchman's report

All the members of the United Nations take note of the IPCC reports and are especially involved when they attend the annual United Nations Climate Change Conference. Most of the delegates are politicians. However, given the IPCC's huge influence, who can check and scrutinise the IPCC? How well-founded are the IPCC's interpretations of what is being observed? How valid are the probabilities given by the IPCC to various climate predictions? The reality could be surprising and for some, even astonishing.

It is important near the end of this short book to check the content of the recent IPCC 2014 Summary Statement (Pachauri et al., 2014).

On track

This 2014 IPCC Summary Statement contains a number of propositions that are discussed in detail. Some of these are obvious and correct and some more true than false. Many could be applied to warming trends in the past:

- It has warmed...
  True, but that also happened in the Minoan, Roman and Medieval Warm Periods.

- The Earth's average temperature rise was around 0.7°C during the past 100 years...
  Probably true, it is hard to know how accurate this 0.7°C estimate is and what is the margin of error in that measurement. In any case the modern warming is similar to the warming in the Medieval Warm Period and average temperatures were higher in the Roman and Minoan Warm Periods and even higher during the Holocene Climatic Optimum over 8000 years ago.

- Glaciers have retreated...
  True, but similar glacial retreat happened in the Minoan, Roman and Medieval Warm Periods and these retreats are well documented, especially in the Swiss Alps.

- Polar ice sheets have lost some mass...
  True, but this would have happened in the Minoan, Roman and Medieval Warm Periods.

- The maximum extent of the Arctic pack ice has decreased in area over recent years...
  True, but this probably happened in the Minoan, Roman and Medieval Warm Periods, and contrary to model predictions the Antarctic pack ice has increased in its areal extent over the past 40 years.

- Sea Level rose 17–21 cm from 1900–2010...
  True, but similar sea level rise would have happened in the Minoan, Roman and Medieval Warm Periods.

- Ocean acidity levels have changed by 1/10 of a pH...
  True, but these words need to be clearly explained to the general public! The wording should be that ocean alkalinity levels have reduced by 1/10 in the pH scale. In this case, **becoming more acid** is better described as **a drop in alkalinity.** The word **acid** can falsely imply a doomsday scenario despite the fact that the oceans have always been alkaline even when carbon dioxide levels were many times higher than those of today. In 2016 the International Council for the Exploration of the Sea, in one

issue of its journal, published 44 peer-reviewed scientific articles critical of many recent ocean acidification studies. For example, some were criticised as being too simplistic, others for ignoring the geological record that shows marine organisms adapt to pH and temperature changes, and others for having an experimental bias that comes from subjecting marine organisms to unrealistic rapid acidification regimes.

· Greenhouse gases such as carbon dioxide, methane and nitrous oxide are at their highest atmospheric levels for 800,000 years...
  Probably true*, but further back in time carbon dioxide levels were up to 15 times higher than today and there was still an ice age.*

· The increases in Carbon dioxide, methane and nitrous oxide gases have been mainly due to industrialisation since the 19th century...
  Probably true*, the new Orbiting Carbon Observatory satellite that is monitoring global carbon dioxide levels will clarify this. It could be that, as the Earth's atmosphere has warmed, more carbon dioxide has been emitted due to the outgassing of carbon dioxide from the world's oceans, and from the oxidisation of vegetation in the tropical rainforests, than first thought. These contributions to atmospheric carbon dioxide levels and methane levels could be just as significant as that from industrial output.*

· The number of warm days and nights has increased and the number of cold days and nights has decreased in mid-latitudes...
  True*, would have happened in any warming period.*

· The warming has increased human heat-related mortality events and decreased cold-related mortality events...
  True*, a truism of any warming period; however, even today, there are more deaths when cold snaps occur.*

· Polar systems are vulnerable to any climate change...
  True*, but polar areas are always very sensitive to climate change. That is not new, nor unprecedented and there could be more adaptive ability by polar wild-life than first thought. For example, polar bear numbers are on the increase, despite frequent statements by Al Gore.*

These comments from the Summary Statements are neither proof of a catastrophic future nor proof that rising greenhouse gases are the primary driver of present climate change.

Off the rails

The IPCC summary statements also contain many interpretative propositions that are either more false than true or just incorrect. Many are even unscientific because they are not based on solid data. These propositions reflect the views of a scientific establishment that has been derailed by obsessively linking rising carbon dioxide levels to a wide range of human catastrophes:

- Sea level rise will continue for centuries because of rising greenhouse gases...
  Speculation, *there have been ice ages and low sea levels with greenhouse levels higher than those of today. There could even be a cooling in the 21st century while greenhouse levels continue to increase and that would stop sea level rise.*

- Each of the three decades since 1980 have been successively warmer than any preceding decade since 1850...
  False, *the first two decades were successively warmer but temperatures have stalled in the latest decade.*

- Multiple lines of evidence indicate a strong, consistent, almost linear relationship between cumulative carbon dioxide levels and projected global temperatures to the year 2100...
  False, *there has not been a linear relationship between temperature and carbon dioxide levels since 1860, as there have been 70 years of pauses in global temperatures in slightly over 150 years. Therefore, there is not a linear relationship at all.*

- It is *'extremely likely'* half the warming has been due to man in the past 50 years...
  Speculation, *if that were the case, then how can there have been a temperature pause since 1998 when carbon dioxide levels have increased over 9%, and a strong temperature rise between 1975 and 1998 when carbon dioxide levels only rose around 6–7%. The correlation between rising carbon dioxide levels and temperature is very poor when longer periods of time are considered! We also do not understand very well the relationship between rising temperatures, the outgassing of carbon dioxide from the world oceans and the oxidation of organic matter in the rainforests. The new Carbon Observatory satellite could resolve this question during the next decade.*

- Man has *'very likely'* contributed to the retreat of glaciers, the melting of ice caps, the loss of ice caps in the past 50 years...
  Speculation, *the contribution must be pretty small, as there has been no change in the long-term sea level rise rate over 160 years. This means the past 50 years are not unique. We cannot measure the loss from ice sheets in Greenland or Antarctica prior to satellite monitoring. We know the warming in West Antarctica for 40 years in the mid-18th century was greater than any warming today. We know the warming*

in Greenland in the 1930s was as strong as in the last decade. The decrease in Arctic pack ice has been more than compensated by the increase in the Antarctic pack ice over the past 40 years. None of these changes is in sync with the linear increase in carbon dioxide.

· There is 'high confidence' from many studies showing negative rather than positive climate impacts on crop yields...
False, there has been a greening of the globe in many areas due to higher carbon dioxide levels and, in general, worldwide agricultural production per unit area has soared due to improved technology.

· Increases in many extreme weather and climate events have been observed since the 1950s...
False, even the IPCC, in the body of the 5th Assessment Report, correctly noted that there were no trends in the annual numbers or frequency of tropical storms, hurricanes, or tropical cyclones over the past 100 years. That report admitted that conclusions in an earlier report, regarding increasing trends in global drought since the 1970s, were probably overstated and that in some countries the frequency of drought had increased, while in others it had decreased.

· Continued emissions will lead to 'severe, pervasive and irreversible impacts' for people and ecosystems...
Speculation, this presumes that global warming is driven by increased greenhouse gases, and overlooks the fact that each increase in carbon dioxide has a lesser and lesser effect. This statement does not differentiate between carbon dioxide (that is non-poisonous and not a pollutant) and industrial gases that are pollutants and a danger to human health. Thus, this statement seems to consider carbon dioxide, a gas most critical to life on Earth, as a pollutant.

· Limiting greenhouse gases can limit climate change risks...
Speculation, rises in greenhouse gases now have a lesser and lesser effect and so any reduction will have a minuscule effect. Even, if true, any reduction would shift things around by an odd year or so in a 100 year-period with no significant change in economic impact.

· Recent changes are unprecedented over 'decades' to 'millennia"...
False, the present temperatures and sea level changes are not that dramatic or unprecedented.

· It is 'virtually certain' that the upper ocean 0–700 metres warmed between 1971 and 2010 and 'likely' warmed' between 1870 and 1971...
Speculation, there are a wide range of opinions. Some say the recent warming is very small. In addition, the resolution of the thermometers used in the deep ocean Argo buoy array of nearly 4000 buoys is not

enough to resolve this statement. The thermometers are only accurate to 0.1°C. There is also evidence the deep abyssal ocean is cooling.

· If we do nothing to limit emissions the Arctic will be free of ice by 2050 'likely', 'medium confidence'…
Speculation, *the latest rebound of Arctic pack ice poses new questions. Al Gore thought the Arctic pack ice would disappear by 2013. There is the possibility of global cooling in the 21st century, and that would throw the cat among the pigeons.*

· Sea level in the period 2081–2100 'very likely' will be in the range 26 cm to 55 cm or with 'medium confidence' between 45 cm and 85 cm if we do not limit greenhouse gases…
Speculation, *sea level has risen at a steady rate for the past 150 years, about 15–17 cm/ 100 years. Without sea level rise getting faster and faster (i.e. accelerating) sea level rise rates well above 15-17 cm/100 years are not just unlikely, they are mathematically impossible. Tide gauges all over the world show the long-term sea level rise rate has been steady over the past 150 years and not got faster and faster.*

· The rate of change will be so fast that most plant species cannot shift their ranges quickly enough to keep up… AND … the rate of change will be so fast that most mammals and freshwater molluscs will not adapt…
Speculation, *in the first place climate change is always faster in sensitive high latitude or high altitude areas and these regions probably always struggle with any climate change. However, as the world warms or cools the rate of temperature change in the mid-latitude and equatorial regions is much lower and there is no evidence that plant species cannot cope with this. James Lovelock puts this clearly when he observes that:*

'The increase of heat occurs mainly at the Poles, and hardly at all at the Equator' (Lovelock, 2014).

· Coral reefs are highly vulnerable to climate change…
False, *coral reefs have always been susceptible to heat waves if a heat wave causes a sudden rise in local ocean temperatures. But reef ecosystems adapt to higher ocean temperatures. Over 8000 years ago, temperatures were much higher during the Holocene Climatic Optimum and sea levels were around 2 metres higher than todays. Coral adapted to these higher temperatures and grew upwards to this higher sea level. Now that sea level is lower than 8000 years ago, some of that older coral is bleached white, and is very, very dead because it stands above present sea level! Sometime tourists on a visit*

to the Australian Barrier Reef are shown this bleached dead coral and the story-line about the bleaching of coral reefs is further exaggerated with dramatic predictions of the demise of the Barrier Reef.

- Global warming is projected to undermine food security...
  False, the opposite is the case. High carbon dioxide levels and warming are helping food production but food supply will be worse if the world cools and the critical wheat belts of the Ukraine, Belarus, Canada and the USA are threatened.

- The situation in the IPCC scenarios RCP 8.5 is considered to compromise human beings and make even working 'outdoors' in high temperatures and humidity a difficult task...
  An unreal, alarmist speculation, this scenario envisages carbon dioxide levels rising by 800 parts per million to 1200 parts per million in the next 85 years when they rose only 120 parts per million in past 100 years. This is a Horsemen of the Apocalypse type projection.

- There are pathways that are 'likely' to limit warming to below 2°C...
  Speculation, the IPCC is preaching a gospel message that reductions in carbon emissions, such as reducing fossil fuel usage, could slow down global warming in the next 100 years. **The situation is that each increase in carbon dioxide now has a minuscule warming effect, as greenhouse warming is near saturation point.** Conversely, this means that any reduction in greenhouse gases will also have a minuscule effect on world temperature. There are valid reasons over time to reduce the use of fossil fuels, as they release pollutants such as sulphur and nitrogen gases into the atmosphere.

It is clear that many of the above interpretations are not scientific statements, but reflect the policies and philosophy of extreme, conservation movements and of rampant climate alarmism.

A word muddle

There are significant uncertainties in climate science. So what level of meaning can be ascribed to the list of terms used throughout the IPCC reports? In order to provide a consistent framework for future probabilities, the IPCC developed a glossary of terms to give an aura of accuracy to its reports. These are:

- confidence: 'very low' to 'medium' to 'very high';
- likelihoods: from 'exceptionally unlikely', 'extremely unlikely', 'very unlikely', 'unlikely', 'about as likely as not', 'more likely than not', 'likely', 'very likely', 'extremely likely' to 'virtually certain'.

Really! How bizarre! How can that level of definition be meaningful, given the nonlinearity of the climate system? How can one take this matrix of terms seriously given the internal limitations of climate modelling and the obvious failure of climate models to predict the pause in world temperatures between 2000 and 2017?

Frozen in time

The initial premises of the IPCC have not changed significantly since its formation in 1988. So many assumptions have not been revised. There is no admission that climate models are in trouble and there is little revision except of the most absurd projections. There is no balanced reporting in the light of recent advances in climate science, nor any serious assessment of the pause in global temperatures since 1998; a factual occurrence clearly evident in the NASA Satellite data. There is no consideration of the possibility of global cooling, despite many peer-reviewed articles about this in the scientific literature. Underlying the IPCC 2014 Summary Statements are only doomsday themes of catastrophic global warming that hark back to the 1980s. All the *old bogeymen* are there, the same patterns of exaggeration, the same old tunes.

No wonder a Professor of Economics at a Sussex University, Professor Richard Tol, resigned from the IPCC in 2014 on the basis that the Summary Statement:

- was the result of encouraging over-estimation due to self-selection of authors and referees.
- omitted improved irrigation and crop yields.
- only showed the impact of sea level rise on the most vulnerable countries.
- emphasised heat stress, but downplayed reduced cold stress.
- warned about mass climate migrations without any solid evidence.
- overestimated the consequences of climate change.
- could have been written by the *Horsemen of the Apocalypse*.

These are damning observations, yet many other similar statements have been made by previous IPCC scientists who have resigned. On climate models a former IPCC author, Dr Steven Japar, who was involved in the 1995 and 2001 IPCC Reports, stated:

Temperature measurements show that the hot zone (that is, a zone predicted by the models in the mid-troposphere) is non-existent. This is more than sufficient to invalidate global climate models and projections made by them.

Some other resignations are worth mentioning:

- Professor Lennart Bengtsson, Director of the Max Planck Institute for Meteorology in Hamburg, a Senior Research Fellow at the University of Reading, from 2008 the Director of the International Space Science Institute in Bern, Switzerland, resigned in 2014 over the extent of intolerance within the climate science community and pressure exerted on him after he joined a right wing group sceptical of some aspects of climate change science. *Environmental Research Letters* refused to publish an article submitted by him that argued for a milder greenhouse effect.

- Dr Christopher Landsea, former Chairman of the American Meteorological Society's Committee on Tropical Meteorology and Tropical Cyclones, resigned in 2005 over the IPCC's exaggerated claims about increased global hurricane frequency due to global warming.

- Professor Paul Reiter, Professor of Medical Entomology at the Pasteur Institute, Paris, resigned in 2001 over IPCC authors trying to link tropical diseases, such as malaria, to global warming.

- Professor Richard Lindzen, Professor of Meteorology at the Massachusetts Institute of Technology between 1983 and 2013, resigned in 2001 on the basis that the IPCC summary statements did not represent what scientists were saying in the larger IPCC technical volumes.

- Professor Roger Pielke Senior, Emeritus Professor of the Department of Atmospheric Sciences at the Colorado State University, resigned in 1995 as he was invited to be a co-author in the 1995 IPCC reports but his comments were ignored, so he resigned.

- Professor Hans von Storch, Professor of Meteorology of the University of Hamburg, was a lead author in the Third Assessment Report (2001), but due to his criticism of Michael Mann's hockey stick description of world temperatures in the past 1000 years (described in Chapter 4 of this book) he was not invited to participate in the 2004 Report.

Conflicts of interest

This short book has examined propositions about present and future climate change promoted by the IPCC. It has not ventured into the business interests and potential conflicts of interests of wealthy leading figures in the climate movement, such as Al Gore, or the previous Chairman of the IPCC, Dr Rajendra Pachauri, or the British economist Lord Nicholas Stern. Pachauri's financial investment arm has been the Tata Energy Research Institute (TERI). Al Gore's financial interests branch out from the Generation Investment Management Limited Liability Partnership and other strategic partnerships. Nicholas Stern's interests are associated with the Grantham Research Institute.

Although there is a basic legitimacy in a free market and the human right to develop business in alternative energy technology and environmental sustainability, it would seem obvious that people with such business interests should not be officials in the IPCC.

The same should be said for scientists on various advisory bodies of the IPCC. In 2009, the Climategate scandal occurred when thousands of emails were leaked to the world media from the computer database of the University of East Anglia. These emails exposed a cartel of scientists keen to exclude scientists with opposing climate views from positions in the IPCC and from publishing in many well known periodicals. Some of these *Climategate scientists* are still dominant figures in IPCC committees.

The IPCC pasta machine

Whatever the quality of some sections in the large IPCC reports, it is clear that even solid, scientific data is processed, as through a pasta machine, and squeezed into a final shape by a clique of individuals so that the IPCC summary statements reflect an exaggerated and false view on how the Earth's climate is behaving. In a United Nation scientific climate report one should not have to sift fact and reasonable argument from propaganda. Climate is chaotic and nonlinear; however, at the present time, there is no evidence of climate behaving outside its normal nonlinear boundaries.

Chapter 14.

Leaving the maze

No big deal

Earth's climate system is a nonlinear system that has survived for nearly five billion years and therefore would have developed significant internal stability. *There is no present greenhouse gas tipping point threatening the Earth.*

The timing of the warming period that began in the early 18th century harmonises with the pattern of past warm periods whose length has varied between 300 and 600 years. However, this warming period is unique as humans have been adding greenhouse gases to the atmosphere since the mid 19th century. This added warming effect, though real, is small.

According to standard physics a curve depicting the warming due to increasing greenhouse gases is *one that keeps flattening out with each increase.* Known as a *decreasing exponential curve*, this is consistent with the data and clearly indicates that *rising greenhouse gases will continue to have a lesser and lesser effect.*

The forecasted rising temperature graphs, used by the IPCC, stem from the high Equilibrium Climate Sensitivity Index used in many climate models. Through this index, water vapour and clouds triple the normal warming greenhouse effect, due to doubling carbon dioxide levels. Yet, in the last 150 years, increasing greenhouse levels have not over-ridden a natural stepwise warming pattern; nor steepened the warming gradient in three warming periods; nor accelerated sea level rise; nor increased the frequency of severe weather. Hence, their effect is properly represented by a decreasing, not increasing, exponential curve.

A lower, not higher, Equilibrium Climate Sensitivity index *is consistent with the data.* Small climate forces that would be overridden by a high Equilibrium Climate Sensitivity index are now significant. Small changes in solar radiation reaching the Earth are apparent. Continual heat exchanges between the Earth's ocean and atmosphere are large enough in degree and timing to be noticeable in the Earth's temperature records.

Global warming is real. Some areas will dry out; climate zones will shift; in some areas bushfires will increase. However, there is no evidence for an acceleration of the Earth's average atmospheric temperature rise or sea level rise. To date there has been no increased frequency of severe weather events. Furthermore, there is also a strong possibility of a cooling period in the 21st century due to changes in the Sun. The quirky nonlinear nature of climate change will continue, as always.

The accelerator bug

Science has caught an *accelerator bug* and is convinced that everything is going faster and faster. It is almost as if the accelerating pace of human development has been transferred to nature. A relevant book on this topic is *Faster: the acceleration of just about everything* (Glieck, 1999). Just because carbon dioxide levels are rising there are false perceptions that sea level rise is accelerating, that severe weather events are more frequent, and that polar ice sheets are on the verge of rapid collapse.

Objective analysis of tide gauge data all around the world shows no long-term acceleration of sea level rise, so the severe sea level rises predicted by the IPCC for the current century are mathematically impossible. Careful analysis of hurricane, tornado, and cyclone frequency reveals no change in frequency of severe weather events. The Greenland Ice Sheet is contributing to sea level rise. However, it is also becoming clearer from satellite analysis that the large Eastern Antarctic Ice Sheet is unlikely to contribute to sea level rise in the 21st century due to extra snowfall in the interior. Such ice gain at present balances ice loss from West Antarctica.

The catastrophic fever

The accelerator bug has so infected the scientific establishment and the world media with the *catastrophic fever* that they tell us that *"this is the critical decade"* and that we are *"on the edge of extinction"* and that everything is going to get worse. Once any type of event is construed as leading to catastrophe there is a witch hunt to punish the progenitors causing extra greenhouse gases; namely, coal, oil and gas. How easy to link the congruence of rising carbon dioxide levels and temperature in the late 20th century and how easy to then predict a doomsday Armageddon?

The simplicity trap

The accelerator bug and the catastrophic fever it causes have been exceptionally virulent and have spawned a host of simplistic arguments. Most sensible people are not aware of the *lack of underlying evidence for increased storm frequency or for accelerating sea levels* and many important people have caught this virus. Prime ministers, national presidents, the current head of the United Nations, two Australian Nobel Prize winners, and even the Pope, are now talking about more severe weather and accelerating sea level rise.

Such talk is a good example of simplistic logic infecting the human mind. Ideas backed by simplistic argument can have a momentum of their own once they are embedded in the popular psyche.

To list a few:

· Thomas Robert Malthus: in 1804 he linked economic growth with a rising population, not realising that as a society became more affluent the birth rate would drop.

· Karl Marx: in the mid 19th century, Karl Marx expounded his simple theories of the need for a class struggle between the upper classes and the proletariat. Revolutionary action by the proletariat would then lead to the sharing of goods and prosperity for all. These simple ideas promoted revolutions, resulting in the death of millions and the replacement of one upper class by another.

· Paul Ehrlich: in the late 20th century Professor Paul Ehrlich of Stanford University equated rising populations with declining food supplies. He predicted imminent global famine because he did not recognise the technological revolution that was occurring in world agriculture. The original edition of his book, *The Population Bomb,* began with this statement:

> *The battle to feed all of humanity is over. In the 1970s hundreds of millions of people will starve to death in spite of any crash programs embarked upon now. At this late date nothing can prevent a substantial increase in the world death rate* (Ehrlich, 1968).

One has only to make a list of climate change book titles to realise just how virulent are the accelerator virus and the catastrophic fever. Journalists are like pods of sharks enjoying a feeding frenzy. Here are some book titles:

FIGHT GLOBAL WARMING NOW

YOU CAN SAVE THE PLANET

SURVIVING CLIMATE CHANGE

WHEN THE PLANET RAGES

THE LONG THAW

IGNITION

THE GLOBAL WARMING SURVIVAL HANDBOOK

HEAT

FIELD NOTES FROM A CATASTROPHE

LIVING IN A WARMER WORLD

HOW CAN WE STOP THE PLANET BURNING?

SIXTH EXTINCTION

A stark perspective

Objective analysis of the cost effectiveness of various carbon dioxide reduction schemes is more than justified. This type of economic analysis is not a subject in this book. For example, strict financial cost-benefit analysis has been performed by a world-famous statistician, Bjorn Lomborg. In general terms Lomborg argues that if carbon dioxide levels projected for the year 2100 are simply reached 10 years earlier after trillions of dollars have been spent, no progress has been made and surely those monies could have been better spent (Lomborg, 2001).

Even Lovelock, as an environmentalist, doubts there is any practical way to significantly reduce the use of fossil fuels quickly enough to control the climate:

> We must not forget the high probability that it is all but impossible for us to reduce the input of fossil-fuel combustion products to the air rapidly enough to change the climate favourably (Lovelock 2014).

The tangled web

The climate debate has been further complicated once greenhouse gases, rising global temperatures, and a number of other issues were all packaged together. This has led to the threat of global warming being manipulated to fight noble causes. So causes, that may have validity in their own right have now latched onto the powerful greenhouse wagon to further promote their case. Opponents of a particular coal mine development, for whatever reason, can now use the greenhouse lobby to fight their battles. Conservative farmers, protesting the intrusion of coal seam gas development on their land, could now encourage the city greens to chain themselves to their farm-gates. Cities fighting serious pollution can now use false arguments that carbon dioxide is a pollutant to help clean up wasteful and pollutant industries that are emitting other gases.

Some sort of *climate politic* has to grace all political platforms, so politicians, whether they like it or not, have to contrive some carbon reduction schemes to get elected. Such schemes have become a pre-requisite for political success. Some are really non-events but look good. Politicians who are reluctant to provide a percentage of a nation's gross national budget to much-needed university research are happy to sponsor visible and costly alternative energy projects so they have something to show the public. A costly experimental tidal plant or wind farm that will eventually become a derelict symbol of economic folly looks good, while the critical funding of our universities and research organisations is seen as a political black hole.

In desperation, scientific and economic research groups have also latched onto the climate debate as a source of funding. In some cases, this has been spectacularly successful. What vice-chancellor or university president is going to knock back any group, if they can bring thousands or even millions of dollars into the university budget?

There is now a large group of university scientists who now form the *'don't-want-to-look-too-closely-brigade'*. These are not climate experts, who are focused on climate science, but if they can give their work some climate flavour they stand to gain that little bit of funding that ensures their survival. This is not to debunk the quality of their research; they are normal opportunists, much as any businessman or entrepreneur.

Sadly, contrary to any tradition of academic freedom, there are other scientists, the *'don't-want-to-raise-their-head-brigade'*, who are experts in climate research and who strongly disagree with the present climate science. In some universities, they cannot say much as they risk employment opportunities and could lose the economic support they provide for their institution and their families.

Co-ordinated attacks on prominent climate sceptics at various universities have been well documented by the media and there are famous scientists who are now not invited to speak at certain institutions. The worst example of such action occurred in September 2015 when 20 signatories, including some IPCC scientists, petitioned the US President to file charges against scientists sceptical of the IPCC position on climate change and carbon reduction schemes by using the Racketeer Influenced and Corrupt Organisations Act (RICO). One could ask with James Lovelock whether political and financial stress has made objective science very difficult:

> I suspect that many climate scientists are aware of the near impossibility of their task. This is what makes me ask: is the pressure from their political pay-masters so great that objective answers to global scientific problems are impossible? (Lovelock, 2014).

Dressed to kill

The rush to install the present alternative energy technologies on a large scale is premature. Widespread installation of solar and wind technology is expensive and actually drains money away from research to develop these technologies further. Even the drive to use more natural gas, instead of oil and coal, is a mixed blessing. Natural gas is the feedstock for fertilisers, plastics and even important pharmaceuticals: chemicals that mankind will need for centuries.

The drive to install infant technology doesn't just come from conservation movements. Clever companies are fuelling the panic. Behind their well presented brochures and slick presentations on how they can save the planet is the secret world of subsidies and kick-backs; often hidden in confidentiality agreements. So many corporations are delighted with their guaranteed incentives.

Smart companies installing alternative technologies can generate a return on capital far in excess of normal risky business enterprises, due to government backed incentive schemes. And with clever packaging, companies installing such technology highlight their environmental responsibility to a gullible public while laughing all the way to the bank. The most bizarre of these schemes proposes to bury carbon dioxide emissions by classifying any extra carbon dioxide produced by industry as a pollutant. What a clever way to forgive man's *guilty behaviour* by classifying the carbon dioxide on which plant life depends as a *poison* and then to get paid to do something that is virtually meaningless!

The banks are also in on this gravy train. For them alternative energy technology projects, the burial of carbon dioxide and carbon trading are prospective new income streams. Like ATM machines, such new business provides a licence to print money as the banks can define monies lent to such projects as risk-capital that requires a higher interest rate even though such projects are often provided with a government-backed guaranteed return on capital.

A death spiral

The drive to install infant technologies infects electrical grids with a virulent cancer. Wind farms, for example, are given the right to sell their electricity even during the night or in quiet periods when their electricity is not needed. This policy means that economical base load coal stations just spin at their own cost as they cannot be shut down. Their only response to protect their economic viability is to raise their base load prices and to shut down some base load units. There is an added concertina effect as troubled grids begin to rely on other regions with surplus power. But then the electricity grids of those nearby regions are placed in jeopardy as they lose their buffer of surplus power. The politicians are now in a quandary! If they remove subsidies, the economics of wind farms will collapse leaving the landscape dotted with graveyards of thousands of rusting windmills. If they don't, the result is a destabilised electrical grid that needs to be rebuilt. In any case vital industries close and relocate to regions that provide economical power. Nothing has been gained as these electrical grids spiral towards disaster. Billions of dollars that could have gone into research to develop more efficient energy technologies to reduce our dependence on oil, coal and gas have been wasted.

A blessing or a curse?

Another climate advocate is appearing late on the scene. The catastrophists have so hammered the idea of the moral imperative for the human race to reduce greenhouse gases or face oblivion, that many religious leaders have decided it is about time they too got into the act, and added their moral force to such argument.

Some religious enthusiasts are urging their followers to make a statement and sell their shares in fossil fuel companies, even though human development over the past two centuries would not have occurred without such fossil fuel companies. However, what good will this grandstanding achieve without adequate alternative energy replacement technology? If these simplistic religious arguments are followed to the letter, the poor will be disadvantaged, third world development will be set back for years and the religious enthusiast will cause the opposite of what was probably intended. Poor under-developed nations need oil, coal and gas immediately and cannot wait for the utopian day when economic alternative energy technologies are developed. That is the reality of the situation.

There are good arguments for philosophies and theologies to be more 'geo-centric', more 'eco-centric' and less 'ego-centric' but there is no justification using them to bolster unproven scientific theories. For example, the Catholic Church, in an effort to be relevant to the modern age, has promoted the climate position of the IPCC. In doing this, the Catholic Church has not learnt from its previous mistakes.

Acceptance by religion of current science is not new but has been often detrimental for the development of science because of the power wielded by religious authorities. Ancient Babylonian science believed the gods created the world in 6 days and had a love-fest in the heavens on the seventh. The writers of the Jewish Bible accepted that cosmology but just inserted one god instead of multiple gods and had God resting on the seventh day. Medieval Christian theology accepted Greek science where the common constituents of matter on the Earth were earth, air, fire and water, while the Moon, Sun and stars were all composed of a fifth perfect unchangeable element.

The Catholic Church has had a mixed run with science when it has tried to reconcile scientific ideas with its theology of the day. This is despite individual clergymen within the Catholic Church advancing science. For example, the Danish Bishop Nicholas Steno was the father of modern geology and the Czech monk Gregor Mendel was the father of modern genetics.

One would think, given the present knowledge of how scientific ideas develop, that modern Churchmen would be wary of getting too close to particular scientific theories. Indeed, backing science consensus may often turn out to be completely wrong and impede scientific progress. As Lovelock remarks:

> I find it depressing that bodies of scientists can talk of their consensus about the extent of climate change when they should all know that a consensus among differing opinions is no answer to a scientific question (Lovelock,2014).

In 1533 Pope Clement VII and some cardinals heard lectures about Copernicus's theory that the Sun was the centre of the solar system. They were very interested and complimentary of his theory. Indeed, Copernicus had an uncle who was a bishop, two sisters nuns, and he himself was a lower order cleric. While some Protestants were objecting, the Cardinal of Capua wrote to Copernicus from Rome entreating him to make his theory more widespread:

> For I had learned that you had not merely mastered the discoveries of the ancient astronomers uncommonly well but had also formulated a new cosmology. In it you maintain that the Earth moves; that the Sun occupies the lowest, and thus the central, place in the universe... Therefore with the utmost earnestness I entreat you, most learned sir, unless I inconvenience you, to communicate this discovery of yours to scholars, and at the earliest possible moment to send me your writings on the sphere of the universe together with the tables and whatever else you have that is relevant to this subject.

Unfortunately, one hundred years later the heliocentric theory of Copernicus was caught up in the Catholic reaction to the spread of Protestant religious movements throughout Europe. So the Catholic Church moved to defend its view that it was the sole interpreter of the Bible and, at the same time, it moved to defend the philosophy of Aristotle that was the basis of Catholic cosmology developed in the Middle Ages. Consequently, the Church condemned the Copernican heliocentric theory and Galileo's own theories about the Earth-like composition of the Stars, the Moon, and other planets in the solar system; a political mistake not rectified for centuries.

Lost in a maze

It is hard to see a way forward in the climate debate due to the inextricable links that have been forged between so many disparate groups. It is like a thousand cats in the same room playing with a thousand balls of wool. It has got to the stage that a sharp global cooling, much like the Dalton Minimum between 1795 and 1820, may be needed to bring the foundations of modern climate science to its knees, and to disentangle the disparate groups caught up in the climate debate.

To some extent we have been conned by our new awareness of the climate in the polar and high altitude regions. We can easily travel to the high latitudes and we are suddenly aware of the dramatic changes that are occurring in those regions in terms of glacial retreat, vegetation change, annual changes in pack ice cover and variations in ice sheet thickness. In September 2015 President Obama had his photo taken near a retreating Alaskan glacier, and used this image to tell the whole world it is on the verge of catastrophic change. *Obama did not tell the world that these polar regions have always been vulnerable to the slightest changes in climate, and always will be.* Climate in the mid-latitudes and the tropics does not have the wild climatic swings of the high polar latitudes.

It is hard to see any way forward unless there is academic freedom and unless scientists are allowed to freely develop new theories. The current widespread lack of academic freedom was exposed by the Climategate scandal when the email database of the University of East Anglia in Britain was leaked to the world media in 2009. This university is a centre of climate research in Britain, and has close ties with the main climate institutions throughout the world. *Yet, here were many emails by prominent climate scientists, in key positions on IPCC committees, discussing how to shut down scientists with different theories on climate and how to stop them publishing in scientific periodicals.*

So now it is not the churches curtailing scientific opinion, but the politicised university world, and a powerful, out of control, IPCC. There is now a *New Inquisition* presided over, not by bishops, but by a clique of scientists who have given themselves the *right of trial* to put scientific heretics to the stake. The new torture methods are not the stake or the rack but the denial of promotion, the manipulation of the media to denigrate, and the refusal to employ. Indeed, efforts to stop such scientists publishing in scientific periodicals have extended to controlling the editorial committees of many well known periodicals. This is the modern equivalent of the Spanish Inquisition's *Burning of the Books*.

The stone age did not end because the world ran out of stone, and, in the same way, there is no reason that the use of coal, oil and gas will end because they are depleted. It will be the inventiveness of the human spirit that will lead the way but that needs academic freedom, an entrepreneurial spirit and the willingness to fund research.

The path forward for alternative energy development is a difficult road because most of the monies will appear to be wasted, even though that is the nature of front-line research. At certain locations, there is a niche for some wind farm technology. However, in many locations the present wind farms are an expensive, wasteful exercise. There has been significant improvement in small-scale solar technology because individual houses in the developed world can afford that technology and that investment has lowered its cost. Unfortunately, there is no large scale solar technology that can economically power large cities, whatever the publicists say.

One of the world's foremost environmentalists adds:

> Huge sums that should have gone on sensible adaptation have been squandered on renewable energy resources, regardless of their inefficiency, or environmental objections, and pointless attempts to achieve that ultimate oxymoron sustainable development (Lovelock, 2014).

The path forward is a difficult one for the conservation movement that is so impatient to curtail the use of fossil fuels. At the present time, the only alternative to fossil fuels is nuclear energy based on uranium fission reactors. It is not surprising that one of the heroes of the global warming lobby, Professor James Hansen, previous head of the NASA Goddard Institute for Space Studies, is a supporter of nuclear energy and the development of fast breeder reactors that use nuclear waste (Hansen, 2009). Hansen is on the outer with the conservation movement on this issue and at the present time the spectre of Chernobyl and Fukushima hangs over the whole nuclear industry. There is considerable research on other types of nuclear energy but the development of fusion reactors, or thorium reactors or reactors using nuclear waste seems years away.

It is also difficult for the conservation movement to admit that, from a statistical viewpoint, reducing carbon dioxide emissions will have a minuscule effect on world temperature, even using the present climate theories. What is the value of spending trillions of dollars so that rising carbon dioxide levels projected for say 2100 will be delayed only 10 years or so? Is not the world still in the same predicament? What would have been the effect if these monies had been spent for the development of the Third World or in pure research?

Whither the IPCC?

The path forward requires a complete restructuring of the IPCC and a more balanced presentation of the climate debate.

In 2005, incredibly, a United Nations Environmental Program prediction of climate refugees showed the ineptitude of the United Nations and its penchant to promote immature and stupid ideas. The prediction of 50 million climate refugees by 2010 derived from questionable calculations by an Oxford University academic, Professor Norman Myers. Of course, there was not one climate refugee in that time. In 2010 the IPCC had to retract a catastrophic forecast on the imminent collapse of the Himalayan glaciers by 2035. In 2013 the temperature forecast for the period 1995–2025 was lowered and that will need to be lowered again. In 2013 the lowest range of the Equilibrium Climate Sensitivity index to a doubling of carbon dioxide was reduced from 2°C to 1.5°C, while the high range of 4°C was not revised; and it is still far too high.

The IPCC needs a complete change of focus so that historical climate patterns are given prime importance over the limited value of computer models. The modellers, sitting at their computer screens, have been so obsessed with modelling and so promoted their complex computing systems, that the IPCC has been driven by this modelling. Historical patterns, even if we cannot explain them, are more important than the models because they reflect and contain the nonlinearity of the system; even though we may not fully understand that nonlinearity.

When will the IPCC:

· Accept that global warming since the middle of the 19th century has been predominantly natural and the warming effects of increasing greenhouse gases an added, secondary ingredient?
· Adopt a balanced position that addresses the possibility of global cooling?

- Accept that the present high sea level rise predictions have been due to a faulty Jason satellite system that needs to be replaced, as there is no coherent agreement between sea level rise data from reliable tide gauges and this satellite?

- Move a motion of gratitude to Mother Earth for all the extra carbon dioxide that is stimulating plant growth and helping Mother Earth cope with the great stress of population growth and a possible population of over 10 billion people by the end of the 21st century?

- Revise its position in the United Nations and tell the assembled delegates not to panic about rising greenhouse gases and to halt the installation of expensive carbon capture schemes and to redirect those funds into energy research?

- Caution poorer nations against installing expensive alternative energy technology schemes based on tax subsidies and tax credits?

Whither science?

What of the broader scientific community? When will it present a more balanced understanding of climate to counter the human propensity to be seduced by overly simplistic logic? Unfortunately, the scientific intelligentsia think they are too smart to be duped by the logic of the day, yet history tells us that false, simplistic ideas have a momentum all of their own and can even dupe the scientific community.

The latest temperature pause for 17 years is Earth's nonlinear climate message to the scientific community. Scientists are now in a very difficult situation. Do they revise their ideas, as John Lovelock has honestly done, or try to save face and maintain the rage?

Some scientific digging has led to arguments that the temperature pause in the first decade of the 21st century is not a pause. Further digging has resulted in global warming suddenly being termed climate change; a sleight of hand way to imply that the reverse is true and that any climate change is somehow related to global warming even if it is global cooling. *Carbon dioxide, the very beneficent gas on which the whole biosphere depends, has been labelled a poison!*

The scientific community will have to throw away a swathe of computer climate models. Such models cannot be used as predictive tools because of the nonlinearity of the climate system. They are also flawed because their internal sensitivity to greenhouse gases in the overall climate system was over-calculated.

There needs to be more detailed examination of climate data from the oceans and from the atmosphere. Historical patterns provide better evidence of how climate behaves than computer models. The detailed data from the new orbiting Carbon Observatory satellite may revise our ideas of how carbon dioxide levels are changing.

We need scientists steeped in nonlinear theory to develop completely new ideas of how this complex nonlinear climate system operates. Despite the raging climate debate, and all the funds being expended, there is actually a shortage of qualified physicists and mathematicians specialising in the detailed physics of climate change at all levels.

In the past 40 years scientists thought that rising greenhouse gases provided a general guide through the climate maze. They were wrong. There are many competing forces behind climate. However, it is going to take some extraordinary events to change the present simplistic carbon dioxide global warming theories within the scientific community.

The fairy tale called the *Emperor's New Clothes* is very relevant. The defence of the role of greenhouse gases as the primary driver of the present global warming may become more strident but one day, someone or some event, will make it clear that the emperor and all his rabid followers are stark naked, and the science community will suddenly nod in agreement and change sides so that climate edifice, built up over the past 40 years, will collapse. Then the charade will be over.

Epilogue

This book has been quite a journey.

When we focused somewhere we were forced to look elsewhere and check many other trails.

We entered scientific and political domains that were blinded by increasing greenhouse gases.

Sadly, neither of these could see how the Sun and the Earth could alter our climate in many different ways.

The journey is on-going, but at least we have escaped from a dark basement into the daylight.

Now there is a chance to travel freely and wonder at the complexity of the climate system.

Selected Bibliography

Akosofu, S., 2010. On the recovery from the Little Ice Age. Natural Science Vol. 2, Issue 11, 1211-1224.

Archibald, D., 2014. The Twilight of Abundance. Why life in the 21st Century will be nasty, brutish and short. Regnery Publishing.

Astrometria – Pulkovo Observatory Internet site. http://www.gao.spb.ru/english/astrometr/index1_eng.html

Baker, R., Flood, P., 2015. The Sun-Earth Connect 3: Lessons from the periodicities of deep time influencing sea-level change and marine extinctions in the geological record. Springer Plus 4:285.

Bar-Server, Y et al., 2012. Geodetic Reference Antenna in Space (GRASP) – A mission to enhance Space Geodesy. (Paper Jet Propulsion Lab. Caltech under contract to NASA).

Bellos, A., 2014. Alex through the Looking-Glass. How life reflects numbers and numbers reflect life. Bloomsbury Publishing. 336pp.

Berner, R., 1990. Atmospheric carbon dioxide levels over time. Science Vol.249. No.4975. 1382-1386.

Bond, G., 2001. A Pervasive Millennial Scale Cycle in the North Atlantic Holocene and Glacial Climates. Science 278.

Broecker, W., 2010. The Great Ocean Conveyor: Discovering the Trigger for Abrupt Climate Change. Princeton University Press 2010. 154pp.

Bruun, P., 1954. Coastal erosion and the development of beach profiles. Beach Erosion Board, US Corps of Engineers.

Bruun, P., 1988. The Bruun Rule of erosion by sea level rise: A discussion of two and three-dimensional uses. Journal of Coastal Research. Vol. 4. No.4. 627–628.

Bryant, E., 2004. Natural Hazards. Cambridge University Press.

Cameron, R., Schussler, M., 2013. No evidence for planetary influence on solar activity. Max-Planck-Institut für Sonnensystemforschung, Max-Planck-Str. 2, 37191 Katlenburg-Lindau, Germany.

Carlson, A., Legrande, A., Oppo, D., Came, R., Schmidt, R., Anslow, F., Licciardi, J., Obbink, E., 2008. Rapid early deglaciation of the Laurentide ice sheet. Nature Geoscience 1: 620–624.

Carter, R., 2010. Climate: The Counter Consensus. Published by Stacey International.

Chen, X., Feng, Y., Huang, N., 2013. Global sea level trend during 1993–2012. Global and Planetary Change. Vol. 112.

Church, J. A., White, N.J., 2006. 20th Century acceleration in global sea-level rise. Geophysical Research Letters. Vol. 33.

Chylek, P., Dubey, M., Lesins, G., 2006. Greenland Warming of 1920–1930 and 1995–2005. Geophysical Research Letters. Vol. 33.

Chylek, P., Dubey, M., McCabe, M., Dozier, J., 2007. Remote sensing of Greenland ice sheet using multispectral near-infra-red and visible radiances. Journal of Geophysical Research. Vol. 112.

Cohn,T., Lins,H., 2005. Nature's Style: Naturally Trendy. Geophysical Research Letters. Vol.32.

Cook, B., et al., 2015. "Unprecedented 21st-Century Drought Risk in the American Southwest and Central Plains," Science Advances Vol 1.

Cook, E., Palmer,J., D'Arrigo, R., 2002. Evidence for a Medieval Warm Period in the 1,100 year tree ring reconstruction of past austral summer temperatures. Geophysical Research Letters Vol.29.

Cooper, J.A.G., Pilkey,O.H., 2004. Sea level rise and shoreline retreat: time to abandon the Bruun Rule. Global and Planetary Change 43. 157–171.

Doe, R., 2006. Extreme floods- a history in a changing climate. Sutton Publishing, England.

Dressler, A., 2010. A determination of cloud feedback from climate variations over the last decade. Science 330, 1523–1527.

Ehrlich, P., 1968. The population bomb. Publisher Random House USA Inc.

Fagan, B., 2000. The Little Ice Age. Basic Books. 246pp.

Fagan, B., 2004. The long summer: how climate changed civilisation. Granta Books, London. ISBN 1862076448.

Fairbridge, R., Shirley, J., 1987. Prolonged minima and the 179-yr. solar inertial cycle. Solar Physics, Vol. 110, 191–210.

Fröhlich, C., 2006. Solar irradiance variability since 1978: revision of the PMOD composite during solar cycle 21. Space Sci. Rev. 125 , 53–65.

Frölich, C., Lean, J. 1998. The Sun's total irradiance: cycles, trends and related climate change uncertainties since 1978. Geophys. Res. Lett. 25 ,

Ge, Q., Zheng, J., Hao, Z., Shao, X.,, Wang, W., Luterbacher, J., 2010. Temperature variation through 2000 years in China: an uncertainty analysis of reconstruction and regional difference. Geophysical Research Letters. Vol. 37.

Gero, J., Turner, D., 2011. Long-Term Trends in Downwelling Spectral Infrared Radiance over the U.S. Southern Great Plains. Journal of Climate, 24, 4831–4843.

Glieck, J., 1987. Chaos – making a new science. Viking Books.

Glieck, J., 1999. Fast – the acceleration of just about anything. Pantheon Books..

Gray, L., et al., 2010. Solar influences on climate. Reviews of Geophysics. Vol 48.

Gribbin, J., Plagemann, S., 1977. The Jupiter Effect. Harper Collins 156pp.

Gribbin, J., 1978. The Climatic Threat: what's wrong with our weather. Fontana Books 206pp.

Haigh, J., 1998. The effects of change in solar ultra-violet emission on climate. American Association for the Advancement of Science. February 1998.

Hansen, J., 2009. Storms of my grandchildren. The truth about coming climate catastrophe and our last chance to save humanity. Bloomsbury Press.

Harig, C., Simons, F., 2015. Accelerated West Antarctic ice mass loss continues to outpace East Antarctic gains. Earth and Planetary Science Letters Vol.415. 134–141.

Hawking, S., 1998. A brief history of time. Bantam Books.

Holgate, S., 2007. On the decadal rates of sea-level change during the twentieth Century. Geophysical Research Letters Vol. 34.

Holzhauser, H., 1997. Fluctuations of the Grosser Aletsch Glacier and the Gorner Glacier during the last 3200 years. In Glacier Fluctuations in the Holocene, ed. Burkhard Frenzel et al., Stuggart. Gustav Fischer Verlag, 1997, 35–38.

Houston, J.R., Dean, R.G., 2011. Sea-level acceleration based on US tide gauges and extensions of previous global-gauge analyses. J. Coast. Res. 27, 409–417.

Houston, J., Dean, R., 2013. Effects of Sea-Level Decadal Variability on Acceleration and Trend Difference. J. Coast. Res. Vol.29. 1062–1072.

IPCC. The reports referred to in this book are in the section Publications and Data at the IPCC website: www.ipcc.ch

IPCC First Assessment Report (1990) (FAR)

Supplementary Report of 1992

IPCC Second Assessment Report 1996 (SAR)

IPCC Third Assessment Report 2001 (TAR)

IPCC Fourth Assessment Report 2007 (AR4)

IPCC Fifth Assessment Report 2014 (AR5)

Special Report on Emissions Scenarios (SRES) (2000)

Special Report on managing the risks of extreme events and disasters to advance climate change adaptation (SREX)

Joughin, I., Smith, B., Medley, B., 2014. Marine ice sheet collapse potentially underway for the Thwaites Basin, West Antarctica. Science. Vol. 344. 735–738.

Kam-biu Liu, Caiming Shen, Kin-sheun Louie., 2001. A 1,000-Year History of Typhoon Landfalls in Guangdong, Southern China, reconstructed from Chinese Historical Documentary Records. Annals of the Association of American Geographers. Vol. 91, No. 3 (Sep., 2001), pp. 453–464

Karlen., W. 2001. Global temperature forced by solar irradiation and greenhouse gases? Ambio. Vol.30. 349–350.

Kench,P., Brander, R., 2006. Response of reef island shorelines to seasonal climate oscillations. Journal of Geophysical Research. Vol. 111.

Kravtsov, S., Wyatt, M.G., Curry, J.A., Tsonis, A.A., 2014. Two contrasting views of multidecadal climate variability in the 20th Century. Geophysical Research Letters. Vol. 41. 6881–6888.

Lamb, H.H., 1991. Historic Storms of the North Sea, British Isles and Northwest Europe. Cambridge University Press.

Le Roy, M., et al., 2015. Calendar-dated glacier variations in the western European Alps during the Neoglacial: The Mer de Glace record, Mont Blanc massif. Quaternary Science Reviews. Vol. 108.

Lewis, N., Curry, J., 2014. The implications for climate sensitivity of AR5 forcing and heat uptake estimates. Climate Dynamics Sept 2014.

Liang, X., Wunsch, C., 2015. Vertical redistribution of Oceanic Heat Content. Journal of Climate. doi: 10.1175/JCLI-D-14-00550.1

Lindzen, R., Choi, Y-S., 2009. On the determination of climate feedbacks from ERBE data. Geophysical Research Letters. Vol. 36.

Lockwood, M., Fröhlich, C., 2008. Recent oppositely directed trends in solar climate forcings and the global mean surface air temperature. II. Different reconstructions of the total solar irradiance variation and dependence on response time scale. Proc. Royal Soc. 464, 1367–1385

Lockwood, M., Owens, M., Barnard, L., Davis, C. and Thomas, S., 2012. Solar cycle 24: What is the Sun up to? Astronomy and Geophysics, 53 (3).

Lockwood, M. and Owens, M. J., 2014. Implications of the recent low solar minimum for the solar wind during the Maunder minimum. The Astrophysical Journal Letters, 781:L7.

Loehle, C., McCullock, J., 2008. A 2,000 year global temperature reproduction based on non-tree ring proxies. Energy and the Environment. Vol.19. No 1. 93–100.

Lomborg, B., 2001. The skeptical environmentalist – measuring the real state of the world. Cambridge University Press.

Lovelock, J., 2014. A rough ride to the future. Publisher – Allen Lane, an imprint of Penguin Books.

Macdougall, D., 2006. Frozen Earth – the once and future story of the ice ages. Published by the University of California Press. ISBN 0 520 24824 4.

Mann, M., Bradley, R., Hughes., Malcolm, K., 1999. Northern Hemisphere temperatures during the past millennium: inferences, uncertainties and limitations. Geophysical Research Letters. Vol. 26(6).

Miskolczi, F., 2010. The stable stationary value of the Earth's global average atmosphere Planck-weighted greenhouse-gas optical thickness. Energy and the Environment Vol. 21. No 4. 2010. 243–262.

Miyahara, H., Yokayama, Y., Yasuhiko, T., 2009. Influence of the Schwabe/Hale cycles on climate change during the Maunder Minimum. Proceedings IAU Symposium No.264.

Monbiot, G., 2006. Heat, how we can stop the planet burning. Allen Lane publishers.

Moran, E., (editor) 2014. Climate change: the facts. McPhersons Publishing.

Muenchow., A., 2014. Internet Site-Icy Seas (http://icyseas.org/2014/11/20/changing-weather-climate-and-drifting-arctic-ocean-sensors).

Pachauri, R., et al., 2014. Climate Change 2014 – a synthesis report. A summary for policy makers. IPCC Publications.

Ollila, A., 2014. The Potency of Carbon Dioxide as a Greenhouse Gas. Development in Earth Science Vol. 2, 2014.

Ollila, A., 2015. Website –www.climatexam.com

Paltridge, G., 2009. The Climate Caper – facts and fallacies of global warming. Connor Court Publishing.

Plimer, I., 2009. Heaven and Earth. Connor Court Publishing.

Price, P., Nagornov, O., Bay, R., Chirkin, D., He, Y., Miocinovic, P., Richards, A., Woschnagg, K., Koci, B., Zagorodnov, V., 2002. Temperature profile for glacial ice at the South Pole: Implications for life in a nearby glacial lake. Vol.99. No.12. 7844–7847.

Przybylak, R., 2000. Temporal and spatial variation of surface air temperature over the period of instrumental observations in the Arctic. International Journal of Climatology 20, 587–614.

Riccobono, F et al., 2014. Oxidation Products of Biogenic Emissions Contribute to Nucleation of Atmospheric Particles. Science. Vol. 344. No 6185. 717–721

Rignot, E., Box, J., Burgess, E., Hanna, E., 2008. Mass balance of the Greenland Ice Sheet 1958–2007. Geophysical Research Letters. Vol.35. Issue 20.

Rignot, E., Mouginot, J., Morlighem. M., Seroussi, H., Scheuchi. B., 2014. Widespread, rapid grounding line retreat of the Pine Island, Thwaites, Smith and Kohler glaciers, West Antarctica, from 1992 to 2011. Geophysical Research Letters. Vol.41. Issue 10. 3502–3509.

Rosenthal, Y., Braddock, L., Oppo, D., 2013. Pacific Ocean heat content during the past 10,000 years. Science Vol. 342, No.6158, 617–621.

Rowlands, D., et al., 2012. Broad range of 2050 warming from an observationally constrained large climate model ensemble. Nature Geoscience 5, 256–260.

Ruppel, C., 2011. Methane hydrates and contemporary climate change. Nature Education Knowledge. 3 (10) 29.

Scafetta, N., 2007. Testing an astronomically based decadal-scale empirical harmonic climate model versus the IPCC (2007) general circulation models. Journal of Atmospheric and Solar-Terrestrial Physics. Vol. 80. 124–137.

Scafetta, N., West, B., 2006. Phenomenological solar contribution to the 1900–2000 surface warming. Geophysical Research Letters. Vol. 33.

Scheuchi, B., Mouginot, J., Rignot, E., 2012. Ice velocity changes in the Ross and Ronne sectors observed using satellite radar data from 1997 and 2009. The Cryosphere, Vol.6. 1019–1030.

Scotese, C. website of palaeotemperature maps. http://www.scotese.com/climate.htm

Shaviv, N., 2003. The spiral structure of the Milky Way, cosmic rays, and ice age epochs on Earth. New Astronomy. 8 (1): 39–77.

Shaviv, N., Prokoph, A., Veizer, J., 2014. Is the solar systems's galactic motion imprinted in the Phanerozoic climate? Scientific Reports 4, No.6150.

Spencer, R., Braswell,W., 2011. On the diagnosis of radiative feedback in the presence of unknown radioactive forcing. Journal of Geophysical Research. Vol.115.

Steinhilber, F., et al. 2009. Total solar irradiance during the Holocene. Geophysical Research Letters. Vol.36.

Steinhilber, F., et al., 2013. 9,400 years of cosmic radiation and solar activity from ice cores and tree rings. National Academy of Sciences Vol. 109. 5967–5971

Svensmark, H., Friis-Christensen, E., 1997. Variation of cosmic ray flux and global cloud coverage – a missing link in solar climate relationships J. Atmos. Sol.-Terr. Phys. 59. 1225ff.

Svensmark, H., 2007. Astronomy and Geophysics Cosmoclimatology: a new theory emerges. Astronomy and Geophysics. Vol. 48(1).

Thomas, E., et al., 2013. A 308 year record of climate variability in West Antarctica, Geophysical Research Letters. Vol. 40.

Thomas, E. R., et al., 2015, Twentieth Century increase in snowfall in coastal West Antarctica., Geophysical Research Letters. Vol. 42.

Tobias, S., Weiss,N., 2000. Resonant Interactions between solar activity and climate. Journal of Climate. Vol. 13.

The Impact Team. 1977. The weather conspiracy: the coming of a new Ice Age. Ballantine Books, 1977.

Tyson. P., Karlen, W., Holgren, K., Heiss, G., 2000. The Little Ice Age and Medieval Warming in South Africa. South African Journal of Science. Vol.96. No.3. 121–126.

Tzedakis, P., et al., 2012. Can we predict the duration of an interglacial? Clim.Past. Vol 8. 1473–1485.

Vecchi, G., Knutson, R. 2015. Historical changes in Atlantic hurricane and tropical storms. Geophysical Fluid Dynamics Laboratory/NOAA report (http://www.gfdl.noaa.gov).

Veizer, J., 2005. Celestial climate driver: a perspective from four billion years of the carbon cycle. Geosciences Canada Vol. 32. No.1. 13–30.

Villalba, R.T., 1994. Tree ring and glacial evidence for the Medieval Warm epoch and Little Ice Age in southern South America. Climate Change. Vol.26. no.203.

Wainwright, D., Lord, D., 2014. South Coast Regional Sea-Level Rise Planning and Policy Response Framework. Draft Report for Shoalhaven City Council and the Eurobodalla Shire New South Wales Australia. (Shoalhaven City Council website).

Watson, P.J., 2011. Is There Evidence Yet of Acceleration in Mean Sea Level Rise around Mainland Australia? Journal of Coastal Research. Vol.27. Issue 2. Vol.368–377.

Webb, A., Kench, P., 2010. The dynamic response of reef islands to sea level rise: evidence from multi-decadal analysis of island change in the central Pacific. Global and Planetary Change. Vol. 72: 234–246.

Wilson, T., et al., (2015) A marine biogenic source of atmospheric ice-nucleating particles. *Nature*, 2015; 525 (7568).

Winkless, N., Browning,I., 1975. Climate and the affairs of men. Publishers, Peter Davies London.

Winter, B., et al., 2006. Extending Greenland temperature records into the late eighteenth Century. Journal of Geophysical Research. Vol. 111.

Wyatt, M.G., Curry, J., 2013. Role for Eurasian Arctic shelf sea ice in a secularly varying hemispheric climate signal during the 20th Century. Climate Dynamics, 2013.

Yu, F., Luo, G., 2014. Effect of solar variations on particle and cloud condensation nuclei. Environmental Research Letters. No.9.

Zarkova, V., 2015. Royal Astronomical National Meeting 2015 – Report 4.

Zelinka, M., Hartmann, D., 2010. Why is longwave cloud feedback positive? Journal of Geophysical Research, Vol. 115.

Ziegler, P., 2013. Mechanisms of Climate Change. http://ktwop.files.wordpress.com/2013/03/climate-change-ziegler–2013.pdf

Zwally, H., et al., 2015. Mass gains of the Antarctic ice sheet exceed losses. *Journal of Glaciology*, 2015 DOI: 10.3189/2015JoG15J071

Figures

Acknowledgements

Figures came from published scientific articles, from the websites of Dr A Ollila; Wikipedia; NASA; NOAA; the British Hadley Centre; the Bureau of Meteorology, Australia. In each case a reference is attached to the diagram. Figures were edited by the graphic artist Juan Carlos Curiel Arroyo. Dominic Brady MSc. and Graham Croker spent hours helping with set out and presentation. Emeritus Profesor Peter Flood made many valuable suggestions.

I pay tribute to Professor Peter Noel-Webb, once Chairman of Geology at Ohio State University and earlier at Northern Illinois University, and to the staff of Northern Illinois University U.S.A. who have been involved in Arctic and Antarctic research for decades. They all encouraged my scientific research in Antarctica. The Alumnus Scientist Award in 2011 from Northern Illinois University US for my contributions to Antarctic research was deeply appreciated. There was also my good friend and supporter, Dr Don Adamson of Macquarie University, Sydney (now deceased). Don, agnostic and humanist, was the eclectic thinker gravitating without any bias to anything that took his fancy from soils, to fossils, to botany, to ancient Egyptian pottery, to supersonic wind on the edge of the Antarctic Plateau. We were close friends and I know Don would love some of the contrary thinking in this book.

I see the quirkiness of the climate system every day I walk along beaches south of Sydney, Australia. The vegetation in the fore-dunes has trapped so much sand that these beaches are stable or even advancing seaward as sea level rises. This behaviour is opposite to that predicted by models showing shorelines retreating as sea level rises. This sort of deviant behaviour is so typical of the climate system. In science, we cannot take any simple argument or model for granted.

About the author

Credentials

Dr Howard Brady has a MSc. from Northern Illinois University, DeKalb, USA and a PhD. from Macquarie University, Sydney, Australia. His research used microscopic fossils to charter the past climatic environments and the geological history of the McMurdo Sound and Ross Sea regions of Antarctica. Dr Howard Brady has published scientific articles in periodicals such as the *Journal of Glaciology, Nature Magazine* and *Science Magazine*. He was a contributor to *Antarctic Geoscience*, a book released by the Scientific Committee on Antarctic Science in 1982. Howard is a member of the Explorers Club of New York and a member of the Australian Microscope and Microanalysis Society. In 2015 Howard was elected a member of the Australian Academy of Forensic Sciences.

Special Award

In 2011, for his contribution to Antarctic research, Northern Illinois University presented Dr Howard Brady with the Distinguished Alumnus Scientist of the Year Award.

> ...*Dr Brady is internationally recognised for his identification and naming of several species of algae currently used to help trace the climate history of Antarctica. Professor Ross Powell observes the continuing influence of Dr Brady's research. The notable Antarctic marine diatom fossil called* Thalassiosira torokina, *which Dr Brady named as part of his M.Sc. thesis, is currently being studied by Professor Reed Scherer's students.*
>
> *In what Dr Brady considers his most innovative paper, he illumined the formation of salts under freezing ice shelves and their subsequent injection like toothpaste on the top of the Ross Ice Shelf (first noted in the early 1900s). Such applications of knowledge in many areas have led to solutions evading others for decades....*

(Dean Christopher K McCord, College of Liberal Arts and Sciences, Northern Illinois University, DeKalb, Illinois. USA. October 11, 2011.)

Made in the USA
Middletown, DE
01 December 2017